RETHINKING CLIMATE CHANGE RESEARCH

Ashgate Studies in Environmental Policy and Practice

Series Editor: Adrian McDonald, University of Leeds, UK

Based on the Avebury Studies in Green Research series, this wide-ranging series still covers all aspects of research into environmental change and development. It will now focus primarily on environmental policy, management and implications (such as effects on agriculture, lifestyle, health etc), and includes both innovative theoretical research and international practical case studies.

Also in the series

A New Agenda for Sustainability
Edited by Kurt Aagaard Nielsen, Bo Elling, Maria Figueroa and Erling Jelsøe
ISBN 978 0 7546 7976 9

At the Margins of Planning
Offshore Wind Farms in the United Kingdom
Stephen A. Jay
ISBN 978 0 7546 7196 1

Contentious Geographies
Environmental Knowledge, Meaning, Scale
Edited by Michael K. Goodman, Maxwell T. Boykoff and Kyle T. Evered
ISBN 978 0 7546 4971 7

Environment and Society
Sustainability, Policy and the Citizen
Stewart Barr
ISBN 978 0 7546 4343 2

Multi-Stakeholder Platforms for Integrated Water Management
Edited by Jeroen Warner
ISBN 978 0 7546 7065 0

Protected Areas and Regional Development in Europe
Towards a New Model for the 21st Century
Edited by Ingo Mose
ISBN 978 0 7546 4801 7

Rethinking Climate Change Research
Clean Technology, Culture and Communication

PERNILLE ALMLUND, PER HOMANN JESPERSEN
and SØREN RIIS
all at Roskilde University, Denmark

ASHGATE

Published by
Ashgate Publishing Limited
Wey Court East
Union Road
Farnham
Surrey, GU9 7PT
England

Ashgate Publishing Company
Suite 420
101 Cherry Street
Burlington
VT 05401-4405
USA

www.ashgate.com

British Library Cataloguing in Publication Data
Rethinking climate change research : clean-technology, culture and communication. – (Ashgate studies in environmental policy and practice) 1. Climatic changes–Research. 2. Climate change mitigation. 3. Environmental protection–Citizen participation. 4. Green technology. 5. Communication in climatology. 6. Mass media and the environment.
I. Series II. Almlund, Pernille. III. Jespersen, Per Homann. IV. Riis, Soren.
363.7'38747–dc23

Library of Congress Cataloging-in-Publication Data
Rethinking climate change research : clean-technology, culture and communication / by Pernille Almlund, Per Homann Jespersen and Sxren Riis, [editors].
    p. cm. — (Ashgate studies in environmental policy and practice)
Includes bibliographical references and index.
ISBN 978-1-4094-2866-4 (hardback : alk. paper) — ISBN 978-1-4094-2867-1 (ebook : alk. paper) 1. Climatic changes—Research. 2. Environmental policy. 3. Culture and communication. 4. Green technology. I. Almlund, Pernille. II. Jespersen, Per Homann. III. Riis, Sxren.
 QC902.9.R47 2012
 577.2'76—dc23

2011053310

ISBN 9781409428664 (hbk)
ISBN 9781409428671 (ebk)

Printed and bound in Great Britain by the MPG Books Group, UK.

Contents

SECTION 3   CULTURE

SECTION 4   COMMUNICATION

List of Figures

List of Tables

List of Contributors

Pernille Almlund is an Associate Professor at the Department of Communication, Business, and Information Technologies at Roskilde University.

Jan Andersen is an Associate Professor at the Department of Environmental, Social and Spatial Change at Roskilde University.

Jørgen Ole Bærenholdt is a Professor of Human Geography and the Head of the Department of Environmental, Social and Spatial Change at Roskilde University.

Araceli Bjarklev is a PhD Student at the Department of Environmental, Social and Spatial Change at Roskilde University.

Thomas Budde Christensen is an Associate Professor at the Department of Environmental, Social and Spatial Change at Roskilde University.

Harry Collins is a Distinguished Research Professor of Sociology and Director of the Centre for the Study of Knowledge, Expertise and Sciences (KES) at Cardiff University.

Anders Danielsen is a project manager at the Section of Development at Roskilde Festival and MA in Communication and Social Science at Roskilde University.

Joshua Forstenzer is a PhD Student in the Department of Philosophy at The University of Sheffield.

Gabriela Ramirez Galindo is an independent communication consultant with more than a decade working experience as a communications professional in environment and development at international and national organizations.

Gert Goeminne is a Postdoctoral Fellow at the Research Foundation – Flanders (FWO) and is affiliated with the Centre Leo Apostel for Interdisciplinary Studies at Vrije Universiteit Brussel and the Centre for Sustainable Development at Ghent University.

Reiner Grundmann Deputy Director of the Aston Centre for Critical Infrastructures and Services (ACCIS) at Aston University.

Ole Erik Hansen is an Associate Professor at the Department of Environmental, Social and Spatial Change at Roskilde University.

Jesper Holm is an Associate Professor at the Department of Environmental, Social and Spatial Change at Roskilde University.

Andrew Jamison is a Professor in the Department of Development and Planning at Aalborg University.

Dale Jamison is a Professor of Environmental Studies and Philosophy, Affiliated Professor of Law and Director of Environmental Studies at New York University.

Per Homann Jespersen is an Associate Professor at the Department of Environmental, Social and Spatial Change at Roskilde University.

Tyge Kjær is an Associate Professor at the Department of Environmental, Social and Spatial Change at Roskilde University.

Corinna Lüthje is a Postdoctoral Researcher at the Institute of Journalism and Communication Studies, University of Hamburg

Rikke Lybæk is an Associate Professor at the Department of Environmental, Social and Spatial Change at Roskilde University.

Paul McIlvenny is a Professor, with a special focus on discourse and society in the English-speaking world, in the Centre for Discourses in Transition (C-DiT) at Aalborg University.

Shameem Mahmud is a Research Fellow at the Institute of Journalism and Communication Studies, University of Hamburg and works in the Adaptive Hazard Management and Climate Change Communication in Asia (HazMan) project of the Clister of Excellence "Integrated Climate System Analysis and Prediction" (CliSAP), University of Hamburg.

Irene Neverla is a Professor at the Institute of Journalism and Communication Studies,University of Hamburg and Principal Investigator of the Cluster of Excellence "Integrated Climate System Analysis and Prediction" (CliSAP), University of Hamburg.

Michel Puech is an Associate Professor in Philosophy at Paris-Sorbonne University.

Markus Rhomberg is an Associate Professor for Political Communication at the Department of Communication & Cultural Management at Zeppelin University Friedrichshafen, Germany.

Søren Riis is an Associate Professor in Philosophy and Theory of Science at the Department of Culture and Identity at Roskilde University.

David I. Schwartz is an Assistant Professor in the Department of Interactive Games and Media at Rochester Institute of Technology.

Thomas P. Seager is an Associate Professor in the Sustainable Engineering & the Built Environment at Arizona State University.

Evan Selinger is an Associate Professor in Philosophy at Rochester Institute of Technology.

Bent Søndergård is an Associate Professor at the Department of Environmental, Social and Spatial Change at Roskilde University.

Susan Spierre is a PhD student in the School of Sustainability at Arizona State University.

Inger Stauning is an Associate Professor at the Department of Environmental, Social and Spatial Change at Roskilde University.

Nico Stehr is the Karl Mannheim Professor of Cultural Studies at the Zeppelin University, Friedrichshafen, Germany.

Acknowledgements

An anthology such as this takes an enormous amount of effort and coordination from a number of different people. We would thus like to thank the following people and institutions for their help and commitment to this project, without them this book would never have been possible.

Thank you to all the authors who have contributed to this anthology, their new ideas and different perspectives were the catalyst for this endeavor.

Thank you to all the internal and external academic referees, whose insight and critique have played an important role in the direction of this book. External referees were Asta S. Nielsen, Christian Kampmann, Dale Jamieson, Erling Jelsøe, Jan Inge Jönhill, Lars Kjerulf Petersen, Michael Søgaard Jørgensen, Ole Jess Olsen and Sara Strandvad.

Thank you to Ashgate, and Katy Crossan in particular, for recognizing the potential of this work and publishing it. This entire production would not have been possible without you.

Thank you to Roskilde University for its support and help throughout this entire process. RUC has been an integral part of brining this project into being.

Thank you to Region Sjaelland for their continued support throughout this project. Their generosity has made this all possible.

And finally, a big thanks to Anna Glasser who has tirelessly worked to see this anthology through to the finish.

Pernille Almlund, Per Homann Jespersen and Søren Riis
Roskilde, Denmark 2011

SECTION 1
Introduction

Introduction to Section 1:
Towards a New Agenda for
Climate Change Research

Pernille Almlund, Per Homann Jespersen and Søren Riis

In many ways the COP-15 climate conference in Copenhagen in December 2009 was a wake-up call regarding our understanding of the complex challenges posed by climate change. Politically, we came to realize that the days of Western hegemony were over and that the European Union's understanding of itself as the avant-garde of the climate movement did not stand the test – the EU was left behind as a spectator during the grand finale of the conference. With regards to the role of science, we had to acknowledge that the opinions and warnings from an almost homogenous scientific community were far from sufficient to generate a plan of action to keep us and our descendants safe from irreversible climate changes.

However, we have also seen a myriad of new initiatives aimed at combating climate change. NGOs, alliances of mayors, business initiatives and not least alliances between local authorities, business and NGOs have taken up the challenge where traditional international politics has more or less failed.

A lot of climate researchers are sitting frustrated – they have put an enormous effort into the IPCC and supporting work and they have reached an almost unanimous agreement among themselves. Their message has been put forward very clearly; there is a 90% risk of arriving at a situation where climate change will be so serious that it will wreak havoc in the social and economic structures of our planet. These scientists' situation can be compared to that of the Trojan princess Cassandra, who was able to foresee future events, but unfortunately nobody took notice of her predictions. That is to say, there is a deep scientific understanding about climate change, which is combined with a powerlessness to impact the current situation.

This combination should serve as a call for more collaboration and inclusiveness amongst researchers from different disciplines and sciences, as well as with the public. If the process to fight irreversible climate change is to be put on track again, climate change and not least climate change politics must be understood in a much broader context – what it means to our quality of life, welfare, social structures etc.

In order to address the challenges, a number of different scientific and public perspectives will need to be integrated. This volume has chosen to focus on three different yet interdependent disciplines, namely the study of: clean technology, culture and communication. Combining these three perspectives is meant to contribute to a broadening of understanding climate change. Examining climate change research through an interdisciplinary approach should help expand how we conceptualize both problems and solutions to the current climate change dilemma. The following volume strives to demonstrate how the combination of clean technology, culture and communication can serve to nuance our understanding of climate change research. With this volume we hope to demonstrate how clean technologies are a necessary part of the solution to climate change, how cultural studies are inevitable if all nations and cultures are to be a part of the solution to climate change and how communication studies play an essential role in providing information and insight into our understanding about climate change from a multitude of perspectives.

It is much too easy for climate sceptics to paint a picture of a future of deprivation, need and misery. A lot of things that you do today will not be possible in the future – a bleak perspective for all those that have got something to lose. Thus, a new and interdisciplinary climate agenda must be outlined. Discussions of mitigation and adaptation to climate change have, in some ways, created a new conceptual framework for addressing problems posed by climate change. While a discussion and differentiation between mitigation and adaptation needs further clarification, which we will return to later in this introduction, the new climate agenda should provide a quantum leap forward – it has to be unrelenting, it is absolutely necessary, it is exciting and it might even be fun.

This new agenda has both a material and a non-material aspect. New technology that can substitute greenhouse gas emitting technologies and improve the quality of life, not least in the developing world, is fundamental for achieving the climate targets. CleanTech is part of the solution but far from all of the solution.

Technology is not developed in a void. Technology is developed in a concrete social and economic context. Technology is developed with a picture of a future society in the minds of the engineers. There is a big difference between the engineer developing an electric car as a necessity and one who develops it with an imagination of a new kind of mobility. There is a big difference for an investor whether he puts his money in a wind power turbine in order to substitute an old coal-fired power station or he invests in a big system of interconnected renewable energy sources potentially solving the need for non-fossil electrical power. The cultural contexts, as well as the ways in which climate changes are communicated are thus crucial to the successful technological development. We need to obtain a much better understanding of the cultural and communicative processes at work in regards to climate change from a multitude of perspectives.

Living with Change

A Summary of Dale Jamieson's Living with Change

There are many reasons why we need to start thinking more conscientiously about living with change. In many ways, the Economist Cameron Hepbrun captures the essence of the concerns surrounding climate change in stating that "If you wanted to design a pill that would put an end to the human race it would produce many of the effects of climate change".

We face a number of great difficulties in dealing with climate change. Some of these difficulties are closely linked to our modern democracies and yet others are due to our cognitive biases. My view of climate change is quite depressing, in that I view climate change not as a challenge to which we are arising, but as a problem that we have already failed to solve. There are however, different degrees of failure: We can fail worse and we can fail less badly. At this stage then, it's time for the post-mortem – it is time to try and figure out why we failed and what we can learn from our mistakes.

We have known about the existence of climate change for a long time. There is a very long history of the study of climate change going all the way back to 1820 with Joseph Fourier, who was really the first to use the analogy of the greenhouse effect. John Tindell, the British Climatologist, was the first to identify the gases that trap heat. Svante Arrhenius was a Swedish Nobel Prize winner who already calculated in 1903 that doubling CO_2 would produce a 1.5–4.5 centigrade warming. In fact, many years later and not to mention millions of dollars later with the development of super computers, we really have not advanced the calculations of Arrhenius to any great extent. It turned out that Arrhenius was largely right, which raises the question that if we had a Nobel Prize winner telling us that doubling CO_2 in the atmosphere was going to produce this result more than a century ago, why do we continue to seem so surprised about this?

A lot was made of the 2007 IPCC report, as if we suddenly had discovered that climate change was real; we had determined the smoking gun. Before this point, it was in some ways reasonable to be a climate change denier, or at least a sceptic, after 2007 however, it appeared that the science was now in and that we should take action. But if you actually look at the IPCC reports, they simply manifest normal science. The reports send a consistent message, one that gets further developed and further embroidered, but in essence they're not dramatically different. So in 1990, the IPCC states a set of certainties, including that human activity which ejects CO_2 into the atmosphere will increase the greenhouse gas effect– resulting on average in an additional warming of the earth's surface.

In this context, it is also important to point to Sloman's research which goes to show that even if we were able to magically stop all greenhouse gas emissions now and stabilize at a relatively unrealistic level, we will still have climate changes for the next millennium. Under the most optimistic scenarios, we still have warming that goes beyond the next millennium; we have CO_2 level rising that will continue to increase over the next thousand years.

I think this is really sobering research. We are used to causing different kinds of harm. If you, for example, stand on someone's foot, and it hurts them and then you stop standing on their foot, the pain dissipates. When it comes to climate change it is as if we have been standing on the foot of nature and future generations and even if we get off the foot now, the pain is going to continue for the next thousand years. That is the world that we have already created.

In my view, the interesting question that we now need to deal with is: Why have we failed to successfully address climate change? It has to do with the way that evolution built us to do a pretty good job at dealing with threats that involve movements of people-sized objects; things like sabre tooth tigers and spears. Evolution did not do a very good job however, at constructing us to deal with threats that are invisible, incremental, have multiple causes and effects, are long term, and/or are probabilistic. Our cognitive systems and our emotional lives are simply not built to really deal with the kind of threat that climate change poses.

We also need to confront the failures of democracy, particularly as they occur in countries like the U.S., which has a strong bipartisanship. Around the globe, and particularly in places like the U.S., people have a very strong bias toward the status quo; accentuating the status quo throughout our political institutions. Democracy falls short in many ways: Politicians have short time horizons, there is a lot of issue competition, the political system does not multi-task very well, and of course, in the U.S. in particular, there is corporate dominance of the political system.

We need to create effective institutions of governance and really challenge the fundamental nature of democracy as it has evolved in the last two centuries. Going back to Jeremy Bentham and other theorists of democracy, it becomes evident that democracies act in the interest of the governed, but they do not govern in the interest of all those who are affected by their effort. Future generations, animals, nature, and those across borders for example, do not have their interests represented in the internal dynamics of democratic societies that are making decisions about climate change. In light of these failures, we need to start rethinking governance, both domestically and internationally.

But most of all we need to learn to cope with change. Change is not the same as sacrifice, although it is often viewed as such. Climate change is already occurring and its rate and intensity will increase and our response to climate change could be a mix of mitigation, planned adaptation, unplanned adaptation, and geo-engineering. I want to emphasize that mitigation is important, without saying, 'don't worry about our emissions just focus entirely on adaptation'. Mitigation matters because it will determine the rate and extent of climate change. But ultimately, and by far the biggest part of the response to climate change, is going to be adaptation, whether planned or unplanned, and increasingly there's going to be a turn toward geo-engineering or at least an attempt to implement forms of geo-engineering.

So my conclusions – my depressing conclusions include: in light of the failure of the COP15, as well as my failed attempt at grief therapy here, that we have failed to successfully address climate change, we are continuing to fail, and we will probably keep failing. The main question now is: How we will cope with our failures and whether or not we will learn from them.

Searching for the Roots and Implications of Climate Change

We may understand culture as sedimented meanings and practices, which a group of people take for granted and thus have stopped questioning. As a consequence, we often do not even see the culture we are embedded in, although it is as an integral part of most of the things we do on a daily basis. From this perspective: traditions, everyday habits and core values are closely connected to the phenomenon we call culture. Examples of pervasive cultural practices are: carrying groceries in plastic bags, taking the car to work and striving to travel far away during holidays. Cultural practices however, may also be expressed by riding a bike whenever possible, recycling waste and using local resources.

The increase of greenhouse gas emissions, spearheaded by industrialized countries, has paved the way for significant global climate change. From a cultural point-of-view, this transformation has not occurred overnight. In fact, the process has been under way for more than a century and is closely tied to industrialization, which is faithfully connected to a definition of the good life in material terms, which reaches even further back in history.

The traditions of production, commercialization, consumption and innovation in the industrialized world have been a self-evident part of life for millions of people in the 20th century. These practices however, are responsible for the advent of climate change. Nevertheless, understanding this connection does not give us clear guidelines for how to deal with climate change, rather to the contrary in fact.

Based on this line of thought, we can better understand that climate change is rooted in our culture. But this also means that the numerous habits and traditions leading to climate change escape our everyday attention, qua culture, which in itself poses a problem. Even more than that, because climate change is so intimately connected to culture, it is with the greatest of difficulty that we may change the causes leading to climate change. It is impossible to fundamentally change the traditions and habits of production, commercialization, consumption and innovation overnight – but if it takes more generations to do so, it will be too late.

In order to deal with the problems connected to climate change, scientists have been good at tracing back climate change to an increase in CO_2 in the atmosphere. Scientists thus argue for a reduction in our carbon footprint. This is an important insight, a century old one which is worth repeating, but in more respects it also equals the now quite banal assessment that links obesity to a relatively high intake of calories. The interesting question – which might also hold some of the answers as to how to live a more sustainable life – is why that person consumes too many calories, despite his or her knowledge of the scientific facts. Analogously to this, it is important to find out why we keep on adhering to traditions of production and consumption, even though they are responsible for an ever-increasing level of CO_2 emissions. This anthology's perspective calls for a more systemic view and comprehensive interdisciplinary teamwork to addressing the challenges posed by climate change. Traditions and habits have a strong hold and momentum, which are often not considered within the natural scientific focus on climate change.

Climate change is symptomatic of an unsuitable culture and lifestyle. If we think we can bypass this connection in dealing with climate change and just focus on CO_2 emissions, then we will most likely simply create other pressing problems. If we only focus on green technologies, failing to see the link between climate change and cultural habits, and just continue our practices and traditions thinking the only thing we need are different "engines" to pull the enormous machinery of the modern world, then we are mistaken. The advent of new technology is important and may postpone climate change, but if we keep up the same rate of production and consumption we have become accustomed to, a number of natural resources will soon be depleted or become so hard to access that it will manifest in the same unsustainable culture that climate change has already wrought on.

It is important for a new agenda on climate change to pay more attention to how climate change is anchored in everyday habits and traditions. It is also just as important to assess and understand how climate change will reconfigure cultures around the globe. We have not yet seen the ultimate consequences of our unprecedently high CO2 emissions. In fact, climate change poses radical challenges not least because it implies a *global* change and thus affects all nations and cultures on earth in one way or the other. Some cultures are likely to disappear as a consequence of climate change, while others may receive thousands of new members. At the same time, some cultures will prosper from the development of new technologies and resilient practices, while others will have to change their means of production. Climate change will rearrange the order of the globe.

To deal with such big changes, common trust and global understanding is needed in combination with a dedicated will to negotiate and seek new solutions. Unfortunately, the COP 15 and COP 16 demonstrated an exact lack of such basic qualities around the globe. As the need for action becomes increasingly urgent however, international institutions and cross cultural collaborations are likely to gain further strength, which may relativize the need for independent nation states. In this way, climate change could come to mark the end of the era of the nation state. To some, this scenario is a positive consequence of climate change and a reason for hope – also to solve a number of other global problems.

However, climate change may also influence cultures the opposite way around, as it may take countries to become increasingly introverted in their effort to adapt to and mitigate climate change. One popular strategy to dealing with climate change is to focus on "the local" and resist global transport and trade, which causes high CO2-emissions. This shift will affect the goals and processes of national and global politics, perhaps even reinforcing local cultures as a consequence. As a result, climate refugees may have a hard time finding new homes, which will give rise to further tensions and maybe even new wars. With stronger self-consciousness and self-awareness however, regions might also develop self-reliance and a new resilience on an effort to take responsibility for their own actions. This approach could also strengthen aglobal cultural (bio-) diversity, making the earth more resilient to rapid changes.

Nobody knows which scenario is going to be fulfilled, if any. But no matter how we deal with climate change, university research on climate change must be strengthened and broadened. As the different perspectives and scenarios outlined above make clear, a new agenda for climate research should be inherently interdisciplinary. Climate change can help decompartmentalize modern research universities and transform them into a unity of concerned investigation.

Inside and Outside Science: Beware of acting too hastily on ClimateGate

Harry Collins

'ClimateGate' refers to the emails hacked from the University of East Anglia's Climate Research Unit at the end of 2009. Suddenly, we were all given a glimpse of the normally hidden, day-to-day workings of a disputed science. There were some unforgiveable aspects, but to a sociologist of science there was nothing shocking about it – it was just business as usual. In spite of the mythology, science cannot get by without humans judging other humans.

One of the most pervasive responses to ClimateGate was, and continues to be, to try to make the science of climate change a more public activity. Thus the insistence of public figures such as, Mike Hulme, Professor of Climate Change in the School of Environmental Sciences at the University of East Anglia, and Jerry Ravetz, Scientist turned Social Commentator and Philosopher of Science, that scientists must 'Show their Working' to the public. In this vain, Hulme and Ravetz write: 'To be validated, knowledge must also be subject to the scrutiny of an extended community of citizens who have legitimate stakes in the significance of what is being claimed'. Here, the two echo the glib, 'one size fits all', contemporary fashion of the social studies of science – 'sort out all problems of visible scientific disagreement by opening things up to the public'.

In the middle of any scientific dispute is a 'core-set' of specialists – these are the people who actually do the experiments, build the theories, and meet together to argue at conferences. In the early days of a scientific debate, such as the one about gravitational wave detection that I have studied in depth, the number of scientists in the core-set could be little more than half-a-dozen, whereas what they do is being reported to and discussed in the outer rings by hundreds of their fellow scientists, funders, policy-makers, journalists, and to some extent, the public at large. The key insight here is that what happens inside the core-set is hugely complicated. In the early 1970s, every waking moment of the scientists locked in the dispute about whether gravitational waves had actually been detected, was filled with calculations, arguments, measurements, judgments of other's capabilities, and so on. And how could it be otherwise? ClimateGate was just a glimpse – a few emails – behind the scenes of a small part of the blooming and buzzing hive of typical scientific controversy.

What it means to be a 'specialist' is to be active within the scientific community throughout the entire 24-hour-a-day buzz. To be a non-specialist means that one is not a part of this environment. If one is on the outside, things inevitably become simplified – the bandwidth is too narrow to carry all the nuanced information about what is happening inside and furthermore, it would be a full-time occupation to absorb it. What happens however is that 'distance lends enchantment'. What is nuanced and unclear to those inside the core-set becomes sharp and clear to those outside of it.

The value system of science is often violated, but it still underpins its distinctiveness. Roughly explained, this means that for the most part, inside the core, scientists are trying to get to the collective truth of a matter, and, when they are serious, they start by trying to understand and fairly represent their opponent's position. This is essential if scientists wish to convince their opponent, as well as themselves, of their own points. And often, those inside the core set have some forlorn grounds to hope that they can convince their opponent with arguments starting from their opponent's position, because they both know the nuances and doubts of all the positions that are at stake. Outside there is no such hope, because no-one knows enough about the nuances and doubts of different positions, so disagreements turn into 'campaigns' rather than debate. This seems a subtle distinction, but it is quite robust and easy to see; scientists immediately know when their opponents have ceased to play by the rules and instead of taking their opponents' arguments seriously, ignore them or caricature them and 'play to the audience'. At this point the scientist is directing argument, not at the core, but outside toward the public. This becomes a 'science war' instead of a science debate. If we want to preserve the thing called 'science' as a distinctive way of making knowledge, we cannot mix the inside too thoroughly with the outside. The movement to 'bring the public into scientific decision-making' is a very good thing, and we can never go back, and should never try to go back. But if things go too far, there won't be any science, there will only be technological populism.

What ClimateGate has shown, apart from some bits of unfortunate sloppy practice, is that the IPCC reports have been mixing the outside with the inside too much. If the IPCC reports had reported solely what went on inside science, there would be more reservations and nuances – the reports would be less useful to politicians. The final political editing stage was what turned them from science to campaign material.

Exploring the Communication of Climate Change

Although we are confronted with a number of communicative challenges, the natural scientific facts of climate change seem to generate the greatest amount of interest in the discussions of climate change. Likewise, political involvement and awareness of climate change, for example via participation in the UN COP-meetings, is still primarily based on the evidence provided by natural scientific facts. When Harry Collins states that the IPCC reports have mixed the inside and the outside of climate science and turned science into campaign material, it is not a turn towards a catch of the communicative challenges, but a stronger focus on scientific facts. Moreover, the media often takes up climate discussions, but primarily focuses on evidence, facts and political will.

With this relatively narrow awareness of climate change based on natural science and political will, we lose sight of communication as an important part of mitigation and adaptation to climate change.

Without communication however, there would be no (public) climate change awareness at all. When NGOs establish international global networks they are based on communication, when politicians, activists, film producers etc. frame and reframe climate change in different ways, they are based on communication and when relatively new technologies such as social medias launch global discussions of climate changes it is communication. Communication frames climate changes and establish discursive formations and this is the reason why we need to follow and analyse communication about climate changes.

As such, climate communication should neither solely be a matter of natural science and fact, nor a case only for journalists and professional communicators.

Climate change should be a matter for communication science, with diverse analyses of climate communication reaching the fields of strategic, political, entertainment, and mediatized communication.

The following section will elaborate upon and discuss the concepts of mitigation and adaptation and will show how communicational and conceptual awareness are important to analyse discussions and decisions about climate change. The concepts 'mitigation' and 'adaptation' have shown to play important roles in the discussions and political decisions about climate change and then play an important role in the communication and understanding of climate change. A conceptual discussion of 'mitigation' and 'adaptation' should contribute to a broadening of the use of the concepts in becoming a part of an interdisciplinary perspective and thereby an element for outlining a new climate change agenda.

Mitigation and Adaptation

The climate crisis occurring in the Arctic provides an obvious point of departure for a discussion about the concepts of mitigation and adaptation, which are both central facets of the scientific and political discussions of climate change. The most recent research conducted on the condition of the Arctic – monitored by The Arctic Monitoring and Assessment Programme (AMAP), which is a working group of the Arctic Council – shows a serious problematic influence of global warming. The ice is irrefutably melting, producing consequences such as a rise in sea level, as well as higher levels of mercury in the Arctic environment due to reemission of the melted ice and ground (AMAP 2011). Despite the publication of this report, discussions within the Arctic Council and the broader political world have still focused on the possibility of more oil and coal exploitation, as well as which of the Arctic Countries can lay claim to territorial property rights due to geological reasons. The climate discussion has been relatively absent and under prioritized in this connection.

Sustaining these priorities will most likely result in more CO_2 emissions, making it impossible to look at the Arctic Council's reaction to climate change as in line with mitigation. Quite the contrary, it could easily be named adaptation, but adaptation that reinforces the damage wrought on by climate change in both the Arctic and elsewhere. In this context, mitigation and adaptation are seen as opposite concepts and strategies; adaptation is understood as an acceptance and adaptation to the current circumstances, while mitigation is understood as an effort to reduce CO_2 emissions and to thereby alter the current circumstances in an endeavour to solve the climate crisis.

In his latest book, "Our Choice – A Plan to Solve the Climate Crisis" Al Gore writes that some of the climate sceptics who reject the scientific consensus and minimize the climate crisis also argue that adaptation is the only possible solution to climate change and that it is not possible to prevent the climate crisis. He goes on to mention that others see adaptation as a dangerous diversion from the serious challenge of solving the climate crisis and taking care of the planet. Al Gore however, finds the differentiation between adaptation and mitigation artificial and writes: "But in reality, this is a false choice. We must undertake both challenges simultaneously, rescuing those in harm's way while saving the future of human civilization. Any other strategy would rightly meet condemnation" (Gore 2009:27).

Many involved in the climate change debate, set up a dichotomy between mitigation and adaptation; understanding mitigation to be a positive strategy in opposition to adaptation, which is viewed as a negative strategy for dealing with climate change. This binary is what Al Gore tries to avoid. Following Gore's line of thinking, adaptation is not necessarily a negative strategy, but in fact can serve as a necessary strategy. Furthermore, we can choose to see adaptation not only in terms of handling the negative consequences of climate change, irrespective of the negative influence of CO_2 emissions, but also as all endeavours of reducing the emission of CO_2. All the different possible actions which can be taken to deal with climate change are adaptations to current circumstances. Who can in fact decide whether integrating wind turbines into the power grid, switching off the lights, or using bio-fuels are techniques of mitigation or adaptation?

In this way, the differentiation between mitigation and adaptation is an artificial one. Adaptation should more precisely be seen as a condition for mitigation, while mitigation efforts can also be thought of as in relationship to adaptation. We adapt to what the world has developed into and to what we expect or want the world to be in the future. In this way, decisions about adaptation are made on the basis of both the past and the future and are fundamental to what is understood as mitigation. Of course not all adaptation is equally good, but the use of adaptation up until now blurs the difference between a diversity of action towards adaptation. Is an increased exploitation of oil and coal from the Arctic as reasonable an adaptation as building dikes in developing countries? These questions should not be answered unambiguously; just as it should be fairly obvious that adaptation cannot be understood as unambiguous either.

Thus, this becomes a call to approach adaptation as a broader concept. Instead of the sole use of an overarching idea about the concept adaptation, we should define what is good or bad adaptation. More precisely, adaptation should be examined on a case-by-case basis, looking at the actions and efforts to see whether they contribute to solving the climate crisis or simply reinforcing it.

Looking more closely at the actions called mitigation, one can see they are understood as actions whose consequences are meant to result in a decrease of CO_2 emissions. This begs the question of what exactly constitutes a decrease in emissions; less emission compared to what level? Is the level compared to the absolute global level of emission or to the local level of a specific action? What are actually the grounds for comparison? At the end of the day, we have to decide which comparisons we will use, for example, we have to collectively decide on an appropriate maximum temperature raise, as well as a global percent of reduction of CO_2 emissions. Unfortunately, these decisions are not often given to the public and even natural scientists underline the complexity and lack of certainty with greenhouse gas reduction schemes. Nevertheless, in order to begin combating climate change we need to make these decisions and accept making them on the basis of probability, otherwise we risk running out of time.

The complexity posed by climate change deserves to be handled with more precise definitions and more caution than we can find in the binary concepts of mitigation and adaptation. However this discussion should not be understood as an effort to remove mitigation and adaptation from the discussion of climate change – rather, it should be seen as a request to use the concepts in a more varied and nuanced way.

The conceptual discussion and use of mitigation and adaptation is one of the reasons to launch a book of this kind, which shows that while natural science is a key science in exploring questions regarding climate change, so too are the social and human sciences. The natural, human and social sciences are combined here in an effort to address the complexity of problems posed by climate change, as well as to broaden conceptual, cultural and technological discussions of climate change. This anthology hopes to demonstrate the importance of this necessity and to broaden the scientific work and solutions of climate change by focusing on clean technology, culture and communication as important factors in the work toward addressing climate change.

SECTION 2
Clean Technology

Introduction to Section 2:
Climate Change and Clean Technology

Per Homann Jespersen

Technology is key to resolving the climate crisis. Without technology, climate policy would only mean the redistribution of fewer resources to an increasing population – a dull society in constant deprivation.

Technology can provide us with the means to substitute scarce resources, to reduce the amount of materials and energy going into our products, to establish more environmentally sustainable ways of fulfilling the needs of mankind.

But it is not a question of just inventing and producing smart devices. As is shown in the chapter by Christensen and Kjær, *What is CleanTech? – Unraveling the buzzword*, technology must be seen in a context of 1) innovation within existing companies 2) support of entrepreneurship to establish new businesses and 3) system innovations transcending the borders of companies and involving multiple stakeholders from business, public authorities, consumers, institutions etc. The introduction of electric passenger cars in Denmark serves as a case to show how innovation on all these three levels is necessary and establishes new business opportunities in the rethinking of the interrelation between the electric power system, technological innovation and the consumer. The technology – the electric passenger car – has been around for more than a hundred years, but to become a commercial success with the potential of eliminating one of the largest and most uncontrollable sources of greenhouse gasses changes on many levels must be undertaken and alliances of stakeholders must be established.

How such alliances might be formed is the theme of the chapter by Holm, Stauning and Søndergaard, *Local transition for low carbon construction and housing – studies of innovative Danish municipalities.* Technologies necessary for reducing the energy consumption of our dwellings and institutions and for substituting fossil energy are readily available and mostly far from 'rocket science' – it is mainly a question of implementing well-known technologies and introducing appropriate standards in new and existing buildings and we will 'immediately' have a substantial and highly cost-effective cut in our greenhouse gas emissions. But despite the apparent simplicity and substantial economic gains when seen from above, things are going much too slowly in the real world.

Holm et al examine a number of best practice cases in which municipalities are leading activities to speed up this process, and analyzing what the essential parameters in these apparent successes are: Building up local competences and institutional capacity in the municipality, establishing networks for market based experimentation, enabling corporations and finances to create a new market on well-known technologies, and finally setting up participatory strategies for the shaping of shared visions. In this way, first-mover municipalities do have the potential for influencing climate strategies far beyond municipal borders.

Along the same lines, albeit in a more global context, Lybæk, Hansen and Andersen's chapter The development of cleaner technologies such as non-fossil energy systems in the absence of strong global governance of climate change develops this theme. Using experiences of the well-known examples of industrial symbiosis in Kalundborg, Denmark and an industrial park situated in Thailand, the authors try to draw conclusions about the successful implementation of climate strategies within the industrial/manufacturing sector when pressure from governments is weak or maybe fully absent – which is evident from the recent COP meetings.

It is stressed that the establishment of a structure of deliberative meta-governance is important for setting up a transition arena, a multi actor network for driving the process towards a higher degree of sustainability, in all of this concept's three dimensions, social, economic and environmental. Examples of successful collaborations – for the companies, for the local community and for the environment – are key to the introduction of a new 'systems understanding' that can inspire other actors and entrepreneurs.

New technologies, in the narrow understanding as 'technical artifacts', are, however, also a part of the CleanTech agenda. In Bjarklev, Andersen and Kjær's chapter, *Possibilities and challenges designing low-carbon-energy technologies – The case of the lighting sector,* alternatives to the well-known incandescent lamp – which in the EU is being phased out – are examined. Observing that a very large amount of the electricity used for lighting is actually used in offices under daylight conditions, this chapter sets out to actually describe how to construct new lighting technologies that may relate to this fact and at the same time reduce energy consumption drastically by using a combination of the most efficient light sources available (LEDs) and daylight led directly into the offices through optical fibers.

Even at this level it is important to take into account several factors. Using the principles of eco-design, it is assured that the product is environmentally friendly not only in the consumption phase but also in the production, distribution and recycling of the used product. Collaboration with industrial designers assures that designs combining modern aesthetics and use value of the lamps are developed. Through interviews with experts and surveys among lay-people different parameters of the product are adjusted in the process of redesigning the way we use and think of lighting.

Finally, Selinger, Seager, Spierre and Schwartz's *Using Sustainability Games to Elicit Moral Hypotheses From Scientists and Engineers* deals with how future developers of technology can learn to deal with the moral dilemmas of including demands for sustainable development into their practice. The understanding of ethics among engineering students is very often based on a number of cases of technology development that has led to disasters. By putting the dilemmas into the form of computer games it is demonstrated how these simplistic views can be enhanced to include social and economic aspects, thus holding a prospect of educating technologists who can cope with the challenges of developing CleanTech.

Together, these papers point to an optimistic CleanTech approach – one that goes beyond constant denial and into a future where caring for the climate can be associated with a belief in technology as something good and helpful based on an understanding of technology in its social and cultural context.

Chapter 2.1

What is CleanTech?
– Unraveling the Buzzword

Thomas Budde Christensen and Tyge Kjær

The Context for CleanTech: New Problems and New Solutions

The idea behind CleanTech arises from the combination of two main phenomena: the human creation of environmental problems and the business opportunities that emerge from attempts to solve these problems.

Environmental problems have a diverse character and cover a broad range of problems and issues. The European commission group environmental problems into four main categories[1]: 1) climate and energy problems related to the emission of greenhouse gasses, 2) problems related to biodiversity and negative effects on nature (air, soil, water, oceans), plants and animals, 3) health related environment problems that cover negative effects on humans following pollution of for example air, food, drinking water etc and 4) resource problems related to the depletion of resource stocks and problems related to recycling issues. The creation of environmental problems varies from industry to industry and often depends on the local geographic and ecological conditions. Eliminating or reducing environmental problems typically involves the creative destruction of existing production practices which consequently give rise to the creation of new business opportunities. CleanTech therefore focuses on the businesses that strategically target these issues in search of new markets and opportunities. In doing so these companies not only contribute to the creation of solutions to environmental problems they also create jobs and economic wealth.

The perception of environmental problems, their solutions and the restrictions and opportunities they pose for society are changing as our society develops. Environmental problems themselves can hardly be characterized as anything new; since the dawn of industrialization, the exploitation and conversion of natural resources into commercial products have generated waste and emissions that have impacted the natural ecosystems. In most countries, industries have for decades been subject to various kinds of restrictions and public legislation aiming at reducing their negative environmental effects. What is new is, therefore, not that industrial activities cause environmental problems and that industries consequently

---

1   This classification of environmental problems is the core of the 6th Environmental Action Programme of the European Commission running for the period 2002–2012

are faced with restrictions but rather that the perception of these restrictions are changing; and that some companies perceive these restrictions as a business opportunity. This new way of looking at environmental regulation was introduced already in the mid-nineties (Porter and Linde 1995; Mol and Jäniche 2009). Today industries increasingly perceive these restrictions not only as limitations and boundaries to their activities but instead as opportunities for new types of activities. This change has during the last decade been amplified by a changing policy approach to major issues; most predominantly to the use of fossil fuels. The political focus on CleanTech which can be found in a vast number of national and EU policies makes it important to understand the concept and its relation to other similar concepts, approaches and policies that have been used during the last twenty years. Furthermore CleanTech is increasingly regarded as an investment opportunity endorsed not only by environmentalists but also by mainstream enterprises and venture capitalists and entrepreneurs. This new development puts CleanTech in a different position in both our societies and economies than it was just ten years ago. This article intends to discuss this new trend and to elaborate on what we can learn from it.

The problem with the concept of CleanTech is that it is an inherently ambiguous concept. It overlaps existing concepts such as cleaner production, eco-efficiency, eco-innovation etc. but yet provides a new angle concerned entrepreneurship and investment opportunities. This element is analyzed in this article.

The article is divided in to two main sections. The first section discusses the definition of the concept 'CleanTech' and analyzes similarities and differences between related concepts. This leads to a typology of three types of CleanTech solutions: single business based CleanTech solutions, lifecycle based CleanTech solutions and system based CleanTech solutions. The second part of the article discusses how new CleanTech solutions are developed and diffused, and elaborates on typical problems that occur in this process. This part also presents a case study concerning the integration of electric car systems in Denmark which illustrates how solutions to environmental problems can create new business opportunities. The case furthermore illustrates how system changes at a sectorial level (in this case illustrated with the energy sector and transport sector) can lead to the creation of new business opportunities.

Defining CleanTech

"CleanTech" often denotes the creation of new business activities on the basis of environmental problems. The following section will discuss what that means.

Pernick and Wilder (2006) define CleanTech in their bestselling book "The clean tech revolution" as "any product, service or process that delivers value using limited or zero non-renewable resources, and/or creates significantly less waste than conventional offerings" (Pernick and Wilder 2006, p.2). They use an inductive approach to unfold the definition by pointing at a wide number of products and services that they characterize as CleanTech. They argue that CleanTech includes four core sectors: energy, transportation, water and materials. The methodology for identifying these exact sectors remains unclear but it is indicated that they are primarily empirically identified on the basis of a study of current investment and market opportunities in the American economy. This empirical approach to identification of CleanTech sectors provides a living and operational concept which from a practical and political point of view illustrates an interesting turn in the American economy, but from an academic point of view appears to provide a coincidental identification of relevant sectors because the principles under which the sectors are identified in the first place remain unclear.

The durability of this definition is illustrated in a later report on job opportunities in the U.S. economy in which the following table over sectors in which job opportunities can be identified is presented:

Energy	Transportation
Renewable energy	Hybrid-electric vehicles
Energy storage	All-electric vehicles
Energy conservation and efficiency	Electric rail
Smart grid devises and networks	Hydrogen fuel cells for transport
Electric transmission and grid infrastructure	Advanced transportation infrastructure
Biomass and sustainable biofuels	Advanced batteries for vehicles
Water	**Materials**
Energy-efficient desalination	Biommicry
UV filtration	Bio-based materials
Reverse osmosis filtration	Reuse and recycling
Membranes	Green building materials
Automated metering and controls	Cradle-to-cradle systems
Water recovery and capture	Green chemistry

Figure 2.1.1 Top CleanTech job sectors

Cooke (2008) proceeds along the same line and argues that a number of sectors qualify as being characterized as CleanTech although the most important sector in the CleanTech economy is considered to be the energy sector.

In a study of the Californian CleanTech cluster, Burtis et al (2004) use a CleanTech concept which is closely related to that of Pernick and Wilder (2006) and define CleanTech as "products and services use technology to compete favorably on price and performance while reducing pollution, waste, and use of natural resources" (Burtis et al 2004, p.11). Like Pernick and Wilder (2006), they emphasize the price and performance in the market-place as a key feature. The definition is used in an analysis of the Californian CleanTech cluster which identifies more or less the same sectors as Pernick and Wilder (2006) but additionally includes agriculture and information industries.

A Danish report about business opportunities for Danish companies engaged in CleanTech defines the concept of CleanTech as "knowledge based products or services that improve reliability, productivity, and efficiency while at the same time reducing costs, energy consumption, waste generation and pollution (Brøndum 2009 page 9, own translation). The focus is in this report more or less identical with the reports and articles mentioned above although the knowledge-based character of the products and services are emphasized in the definition as well as in the following analysis.

Clean Technology

What we can learn from these attempts to define the CleanTech concept is that CleanTech is technologies (or technology systems) that have a specific performance that makes them competitive while at the same time making them less polluting, more resource effective, less waste generating or less hazardous to humans and ecological systems than competing technologies. The concept consists of two basic elements:

Technology: The word 'technology' basically refers to the utilization of knowledge and science to produce commercial products. Technology has therefore historically been closely related the discipline of engineering although today almost all other academic disciplines to some extent have been exploring, analyzing and investigating the concept of technology. Technology is therefore a broad concept that when used in a narrow sense refers to the use of knowledge and hardware to solve problems and when used in a broader meaning includes the complete system of knowledge creation and knowledge exchange between research, industry and society at large. The narrow meaning of the technology-concept covers elements such as process techniques, equipment, machinery etc. but also the organization and management of these including management systems, production layout, quality systems (TQM), logistical systems etc. Technology should therefore not to be confused with hardware only, even when used in the narrowest definition. The broader definition of technology includes the system in which a production techniques and systems are embedded. This wider definition of technology is for example used by researchers that analyze how technology systems (Geels 2005), innovation systems (Lundvall 1992) or clusters (Porter 1998) invent, develop, use, adapt and diffuse technologies.

Technology is furthermore used to describe a specific feature of products or services. Some products that are produced in advanced production processes or products that themselves are regarded to be complex and advanced are often characterized as high-tech products. Also services that employ or utilize advanced techniques or processes are characterized as technology. In between products and services lies the information and communication technology which sometimes takes the form of a product and sometimes as a service. Information and communication is characterized as technology based on the fact that modes and techniques used when communicating and exchanging information is considered to be advanced and thus makes the systems that enables information flow to be characterized as technology. The ambiguities in defining technology and especially high-technologies are important because politicians tend to favor support for high-tech over low-tech solutions as high-tech often is associated higher income jobs and therefore they attract more political attention.

Clean: The other part of the concept CleanTech refers to the performance of the technology or technology system emphasizing that the technology needs to perform in a certain way that makes it less polluting, more resource effective, less waste generating or less hazardous to humans and ecological systems than competing technologies or technology systems. In other words, the notion "clean" is typically used in a relative sense which means that the technology or technology system should be cleaner than competing technologies or systems without in absolute and universal terms defining when something is clean or not. This means from a product point of view that the product itself can be defined as CleanTech if it meets the performance criteria but also that a product can be characterized as CleanTech if it is a part of a technology system that collectively meets the criteria. A light weight material in a car can, for example, be characterized as CleanTech if it contributes to reducing fuel consumption but may fall out of the category if it is used in a sports car that consumes more fuel than conventional cars. Defining whether a product can be characterized as CleanTech thereby often depends on the context and the system the product is a part of.

The relative feature of the concept can lead to ambiguities as some technologies or technology systems may simultaneously solve some problems while creating new ones. A diesel car is for example more energy efficient than a petrol car but cause more particulate emissions. Defining which of the technologies that is cleaner therefore depends on how fuel consumption and particulate air pollution is weighed against each other.

CleanTech and Other Related Concepts

The concept of CleanTech is very closely related to other concepts that have been used over the last thirty years: Some of these are *eco-innovation* and *cleaner production* and the product oriented concepts *eco-friendly technology* and *environmentally sound technologies*. These concepts have been developed and used by private enterprises and organizations such as the UN, the OECD and the European Union in order to promote the development and diffusion of less polluting technology.

Academic literature (Kemp 1997, Berkel 2000, Baas 2005) often distinguishes between two overall categories of environmental technologies: end of pipe measures and cleaner production measures.

End of Pipe Measures

End-of-pipe measures are defined as a set of measures that aim at treating pollutants at the end of a process (UNEP 1996). End-of-pipe measures are therefore most often add-on technologies which are attached to an existing system at the facility or at a communal level without changing the fundamental features of this system (Baas 2005). End-of-pipe measures include technologies like catalytic converters, filters, scrubbers, wastewater treatment facilities etc. End-of-pipe measures are usually considered less effective than cleaner production measures (Remmen and Thrane 2007). The end-of-pipe technologies furthermore also add costs to products due to additional maintenance and repair activities and additional costs in end of life phases (UNEP 2000, Kuehr 2007).

Cleaner Production

Cleaner production is pro-active, integrated and preventative approach focused on the elimination of the environmental impact at the source instead of treating the problem after it has occurred (which is the case with end-of-pipe measures) (UNEP 1998).

The concept of cleaner production was initially developed in the late 1980s when institutions such as the OECD and United Nations Industrial Development Organization (UNIDO) and ministries of environment from various countries (like Austria, Denmark, the Netherlands, Germany and the UK) were active in developing tools and procedures to assist companies reducing their environmental impact (Baas 2005). The approach was further developed in the following years by many different organizations: The International Standard Organization developed the environmental management system (ISO 14000 series), the World Business Council for Sustainable Development developed the concept called eco-efficiency in order to assist companies in linking cleaner production initiatives to the value creation in businesses (WBCSD 2000), the European Commission developed the EMAS environmental management system and implemented the preventative approach in the IPPC directive (Directive 2008/1/EC) that determines how environmental permits should be issued in EU member states. The United Nations Environmental Programme (UNEP) and UNIDO initiated a number of initiatives in order to disseminate knowledge about cleaner production and assist companies in greening their business models, such as the establishment of 45 cleaner production centers around the globe (UN DESA 2007).

Cleaner production is defined in the International Declaration on Cleaner Production by the UNEP (1998) as:

> "The continuous application of an integrated, preventive strategy applied to processes, products and services in pursuit of economic, social, health, safety and environmental benefits." (UNEP 1998)

Cleaner production includes measures that influence the environmental performance of a given system and explicitly aims at pollution prevention rather than clean-up and post-treatment measures (Thrane and Remmen 2007, US EPA 1988). Examples of cleaner production measures are: improving energy efficiency, de-materialization, waste minimization, raw material substitution and changing organizational measures. Because cleaner production measures focus on eliminating the activities that cause the problems they are often regarded to be more effective than end-of-pipe measures and also more cost-efficient in the long run (UNEP 2000, Kuehr 2007).

The OECD has, similar to UNEP, adopted the concept of cleaner production and uses a more extensive definition:

> 1. Cleaner production approaches includes hardware (goods, services, equipment) and software (technical know-how, organisational and managerial skills and procedures). Technology co-operation in support of cleaner production, therefore, can encompass a variety of relevant activities including for instance industrial development, institutional capacity development, education, and policy dialogue. 2. Compared with standard methods, cleaner production techniques and technologies use energy, raw materials and other inputs material more efficiently; produce less waste, facilitate recycling and reusing resources and handle residual wastes in a more acceptable manner. They also generate less harmful pollutants and can assist in abating greenhouse gas emissions. Cleaner production methods have significant financial and economic advantages as well as environmental benefits at the local and global level. Support for cleaner production can thus play an important role in relation to the Framework Convention on Climate Change. (OECD 2000)

Like the UNEP definition, the OECD definition of cleaner production differs from end of pipe measures firstly by emphasizing preventive measures and secondly by including a boarder variety of measures that also embrace activities and factors that lies beyond company boundaries such as industrial development, institutional capacity development, education and policy dialogue.

The cleaner production approach has many features that are associated with the concept of CleanTech such as the idea of identifying activities where companies can profit from reducing their resource consumption, environmental impact and waste generation. However the cleaner production approach mainly targets established industries and provides tools, procedures and management systems that can assist established companies in achieving such targets including the pertinent regulatory and legal systems. CleanTech initiatives as presented by for example Pernick and Wilder (2006), Burtis et al (2004) and Cooke (2008) on the other hand include these issues but tend to emphasize entrepreneurship, venture capital opportunities and new products and markets opportunities over activities that take place within existing industries.

Eco-innovation

Eco-innovation is a concept that is used by the European Commission. Eco-innovation focuses on the process through which environmental technologies are developing, demonstrating, marketing and diffused. It is the central concept in the European Environmental Technology Action Plan (ETAP), which was launched in 2004 in order to promote and assist development and the adoption of environmental technologies to the combined benefit of the environment, economic competitiveness and economic growth. The ETAP is the third pillar of the Lisbon strategy (European Commission 2004a); the first and second pillars focus on improving the competitiveness for the European economy and securing sustainable economic growth with more and better jobs and social cohesion, respectively.

The ETAP (European Commission 2004b) defines *eco-friendly technology* as technologies whose use is less environmental harmful than relevant alternatives – very similar to the way that CleanTech is usually defined. It embraces a broad variety of technologies and processes that can be deployed in order to manage pollution; it covers less polluting and less resource-intensive products and services and it include management systems used to manage resources more efficiently (European Commission 2004b). The definition of *eco-friendly technology* used by the European Commission is based on the UN Agenda 21 definition of *environmentally sound technologies* developed by the United Nations Conference on Environment and Development (UNCED), following the Rio conference in 1992, which states that:

> "Environmentally sound technologies protect the environment, are less polluting, use all resources in a more sustainable manner, recycle more of their wastes and products, and handle residual wastes in a more acceptable manner than the technologies for which they were substitutes.
>
> Environmentally sound technologies in the context of pollution are "process and product technologies" that generate low or no waste, for the prevention of pollution. They also cover "end of the pipe" technologies for treatment of pollution after it has been generated." (UNCED 1992)

The philosophy in the action plan is to build knowledge-based growth and economic prosperity on opportunities that arise when industries and energy systems are transformed in the attempt to solve the environmental challenges that the European economies are facing. Eco-innovations are in other words products, services or technologies that solve environmental problems while at the same time create sustainable economic growth and employment. As such it is very closely related to the idea about CleanTech.

The Scope of CleanTech Solutions

The UN definition furthermore states that environmentally sound technologies not only include individual technologies but embrace the total system of interlinked technologies and services, capacity, knowledge, know-how, routines and skills that surround a given technology (UNCED 1992). This means that a production system like organic farming (that includes a variety of practices and procedures) also qualifies as an environmentally sound technology.

The system perspective is very important as some solutions to environmental problems are more effectively solved if a larger number of companies, consumers and authorities are involved in developing and implementing the solutions. This is also why the eco-efficiency tool developed by the World Business Council for Sustainable Development (2000) encourages companies to include suppliers and consumers when developing working with issues such as reducing material or energy intensity of products, enhancing recyclability, maximizing the use of renewable, extending product durability or increasing service intensity of products.

Creating CleanTech solutions to environmental problems inside single companies additionally runs the risk of creating burden-shifting where the problem is not eliminated but instead moved to somewhere else. This is why a lifecycle approach in which environmental impacts from all phases of a products lifecycle are evaluated often is a preferred way of managing the development of CleanTech solutions.

It is possible to distinguish between at least four different types of burden shifting: 1) Burden shifting can take place between companies in a value chain if, for example, a product change transfers the problem from a lead company to its supplier, 2) it can occur as burden shifting between impact categories, where reduction of one environmental problem leads to the creation of another one, 3) burden shifting between geographical areas for example where a specific problem is transferred to a production unit in another part of the world, or 4) burden shifting between various stages of a products lifecycle, for example where a product is designed to consume less energy in the use phase while at the same time consumes more material in the production. The UNEP and Society of Environmental Toxicology and Chemistry (SETAC) have for that reason developed a set of management tools to assist enterprises that want to integrate a lifecycle perspective in the way to deal with environmental problems in their organizations (UNEP/SETAC 2009) and the International Standardization Organization (ISO) have developed a standard on how to conduct lifecycle analysis called the ISO 14040 standard.

There are, however, also limits to lifecycle based CleanTech solutions. Some environmental problems cannot be solved by initiatives taken by stakeholders associated with one product's or service's lifecycle. A wider system approach may therefore sometimes be needed in order to facilitate changes on a system level. Greening the transport sector with the implementation of electric cars on behalf of combustion engine cars for example requires changes in the car industry in terms of setting up the proper production system capable of designing and producing the electric cars. But changes are also needed in the energy sector that has to deliver electricity produced from renewable energy sources and the charging infrastructure needs to be installed. Such a system cannot be established by companies and consumers linked to a single product's lifecycle.

It is therefore possible to distinguish between three different types of CleanTech solutions:

- **Single company CleanTech solutions:** This type of CleanTech solution arises from initiative taken inside existing companies typically as research and development programmes carried out independently or in collaboration with other research institutions. This type is therefore often internal solutions that may have an external relation. This type of CleanTech can lead to spin-off's from the existing companies or research institutes and thereby result in entrepreneurship and new business creation.
- **Lifecycle based CleanTech solutions:** These types of CleanTech solutions include more than one company and require collaboration between multiple stakeholders involved in various phases of the lifecycle of a product or service. These are in the simplest form organized as collaboration between companies and suppliers or customers in order to develop solutions to environmental problems. More sophisticated types of CleanTech solutions in this category include new business models that arise from changed responsibilities between stakeholders in the lifecycle of a product or service. An example is the so-called "product service systems" in which a company sells a service instead of a material product, for example where a supplier of chemicals agrees to supplying the service that the chemicals create instead of a certain amount of chemicals which switch the incentive structure at the supplier from an incentive to sell more chemicals to an incentive to improve performance and reduce the use of chemicals. Another example is the so-called Energy Service Company systems (ESCO's) where an energy company optimizes the costumer's energy consumption through the application of energy saving technologies and finances the investments through the reduction in the costumer's energy bill (COWI 2008). These examples are all new types of business models that arise from changes in the organization of the value chains. They can hardly be identified if the opportunities are not seen in a lifecycle perspective.

The lifecycle perspective has also been integrated into policy – in the EU the lifecycle perspective has especially in directives under the Sustainable Consumption and Production umbrella (European Commission 2008).

- **System based CleanTech solutions:** System innovation is characterized by CleanTech innovation that transcends the borders of a particular product's or service's lifecycle. It often relies on multiple and simultaneous initiatives taken by various stakeholders from businesses, public authorities, consumers, research institutions etc. Changing energy systems from fossil fuel to renewable energy sources, for example, requires that research institution develop appropriate technologies, that energy companies change investment patterns, that public authorities set up feed-in tariff's system to support diffusion of renewable energy technologies, that transmission grid operators develops solutions that can level out fluctuations in both consumption and production of energy and that consumers accept and demand new price structures.

System based changes are also likely to result in larger and deeper changes to the way the economy is organized which eventually will lead to the loss of assets at the companies that are not able to adjust to the changed environment but at the same time also create a larger window of opportunities for entrepreneurship and new business models.

The Valley of Death

An important discussion revolves around identifying the factors that determine success and failure of new CleanTech products. A typical problem that CleanTech is faced with is bridging the gap between initial basic research and commercialization. The European Union and its member states have invested huge funds in developing technologies that can reduce or eliminate the environmental problems that society is facing, however investments into basic research often strand before they are converted into viable business models. This is a commonly known problem for most types of innovation processes – also those that have little to do with CleanTech. What is special about CleanTech is it focus on businesses, investors, researchers and other stakeholders that strategically target new markets and opportunities that arise from increasing environmental concern in terms of 1) climate and energy related problems, 2) problems related to biodiversity and negative effects on nature, 3) health related environmental problems following pollution, and 4) resource problems related to depletion of stocks and problems related to recycling.

A typical problem associated with large scale research and development programmes concerns bridging the gap between public and private funding that can link the initial technical discoveries to the early commercialization and ensure that new technologies are brought onto the market by private enterprises. The gap between public sector funding and private sector funding is illustrated as the 'Valley of Death' in the figure below. Early entrepreneurs (new businesses or development project within established companies) often find themselves in a high-risk situation where they have to meet public goals and criteria to access funds and at the same time need to meet the demands from the private sector such as profits and return on investments (Murphy and Edwards 2003). The scaling up of laboratory tests to demonstration and pilot plants are furthermore often very costly and time consuming.

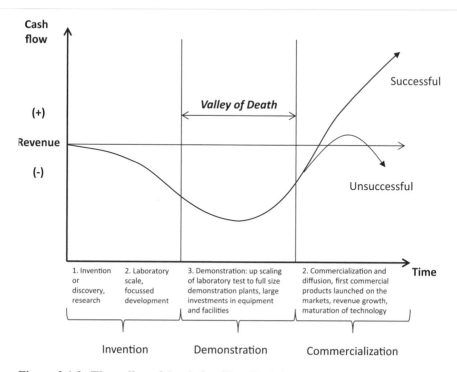

Figure 2.1.2 The valley of death for CleanTech investments

This figure is reprinted and edited with permission from the National Renewable Energy Laboratory (NREL) Report (NREL/MP-720-34036, page 3) titled, "Bridging the Valley of Death: Transitioning from Public to Private Sector Financing" (May 2003) by L.M. Murphy, http://www.nrel.gov/docs/gen/fy03/34036.pdf.

The gap can be bridged in a number of ways depending on the specific circumstances for the given technology. Many renewable energy technologies face a market situation where they are not competitive compared to technologies based on conventional (fossil) fuel sources. This fact raises uncertainty and deepens the 'valley of Death'.

Another feature which is common for CleanTech is that the market conditions are heavily influenced by public sector interventions. Renewable energy technologies often depend on feed-in tariffs, quota systems, tax exemptions or other incentive schemes in order to be competitive to fossil fuel based energy technologies. In fact most renewable energy technologies such as wind turbines, photovoltaics, biofuels etc. are not competitive at all if they were not economically supported by the public sector. Feed-in tariffs can make the slope of the right side of the cash flow line in the model above steeper and consequently make CleanTech more profitable in the early commercialization phase. However the market based incentive schemes are not always enough to bridge the gap and the fact that public support schemes my change due to elections may further raise uncertainty especially when the development programmes runs for more than an election period.

The gap between the invention phase and early commercialization is usually covered by investments placed by entrepreneurs or by seed capital. Venture capitalists typically focus on the following phase when the first sales have demonstrated a future market potential for the technology (Murphy and Edwards 2003). Venture capital is usually invested just before the curve begins to climb and drawn out again when a stable business is established and conventional capital can be attracted in order expand the business.

CleanTech Compared to Related Concepts

The concept of CleanTech bears many similarities with existing approaches, tools and concepts, most notably the focus on exploiting the opportunities for generating a positive economic benefit (to companies and the societies at large) while reducing environmental impacts. However, there are also differences between the various concepts. The differences have to do with how aspects such as product development, process development and entrepreneurship are incorporated and treated. The figure below illustrates some of the similarities and differences between various concepts. The figure distinguishes on the one axis between a focus on established industries versus focus on entrepreneurship and new business models and on the other axis between production processes and operational aspects versus new products and services.

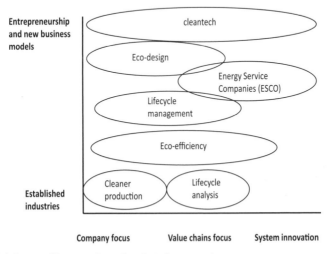

Figure 2.1.3   Cleantech and related concepts

The figure illustrates that the CleanTech concept has a strong emphasis on entrepreneurship and new business models and at the same time focuses on new products and services. The concept is in that respect contrasts the concept of cleaner production that mainly targets established industries and mainly focuses on how existing production processes can be optimized and improved in order to eliminate or reduce the environmental impact.

Case Study: Implementation of Electric Cars

The following case study illustrates how climate change mitigation by the integration of wind turbines in the Danish energy system combined with the introduction of electric cars opens opportunities for new CleanTech business models. The case study is chosen as an example of how challenges to energy systems (integration of fluctuating wind power in the energy system) can be converted into business opportunities. The case thereby illustrates how the system perspective makes it possible to identify new business models and opportunities that may appear invisible from the perspective of the established companies and their value chains. Examples being energy companies (mainly engaged with the production of electricity) or vehicle manufacturers (mainly engaged with the production of cars).

Transport is a vital part of our modern society but the extensive use of transport comes with a high cost to society. Conventional mass-market vehicles consume large quantities of fossil fuel. This causes air pollution (especially in urban areas exposed to heavy traffic) and emits CO_2, which contributes to climate change. Greenhouse gas emissions from the transport sector in Europe grew between 1990 and 2006 by 28% compared, to a reduction of 3% across all sectors (European Environmental Agency 2008). The air quality in several urban areas furthermore failed to meet limit values set by the European Commission (European Environmental Agency 2008).

The problem with increasing CO_2 emissions, poor urban air quality and depletion of fossil fuel reserves provides an opportunity for companies that provide *alternative* technology and transport solutions that can reduce environmental impact. Electric cars transport is one potential solution to the problems. Electric cars benefit the environment in two different ways: First, they have no exhaust pipe emissions and therefore cause no local air pollution when driven. Instead, the pollution is transferred to the energy production sector, where it might be to eliminate emissions altogether by producing the energy from renewable energy sources, or to exploit economies of scale in fossil fuel energy production that makes it easier and cheaper to clean emissions. Second, electric cars are far more energy efficient than conventional combustion-engine cars (Åhman 2001). Introducing electric cars on behalf of conventional combustion-engine cars will thereby reduce energy consumption but also lead to significant greenhouse gas emission reductions as fossil fuelled power stations are replaced by renewable energy sources.

There are a number of reasons why conventional combustion engine cars have not yet been replaced with electric cars despite the environmental benefits. The two main reasons are that the infrastructure is not in place and that the battery technology is still immature which leads to high purchase prices on electric cars. Batteries do not quite offer enough energy to make electric cars competitive with conventional cars. The lacking energy storage capacity in batteries provides electric cars with a limited range and a low top-speed compared to combustion engine cars. Furthermore, the price of batteries is still relatively high, typically $10,000 to $20,000 to the vehicle price compared to a conventional car (PricewaterhouseCoopers 2009). Production plans for electric cars collected by the International Energy Agency (2011) reviled that the car manufacturers only plan to produce around 1.4 million electric cars by 2020 compared to a global total sales volume exceeding 50 million cars. In other words, the electric cars are still located in the valley of death where investments have been placed in order to develop the technology. A large scale commercial break-through is still unachieved. Part of the solution is therefore to identify the business model that will enable vehicle manufacturers to bridge the valley of death.

Integrating Electric Transport and Renewable Energy Systems

The benefits of electric transport depend to a large extent on how clean and efficient the energy system that generates the electricity is. The advantage of electric transport increases when the share of renewable energy sources in the energy production is increased as coal and gas fired power stations are substituted with wind turbines, photovoltaics, hydro power and biomass incineration. Introducing renewable energy into energy systems is often complicated because renewable energy sources tend to fluctuate due to seasons and weather conditions. The figure below illustrates the differences between four types of renewable energy sources. The energy sources differ in terms of variability: both wind, solar and hydro power has a high variability due to seasonal changes and weather conditions. The variability is highest for wind and lowest for biomass. They also differ in terms of predictability: It is possible to predict both biomass production and hydro power but more difficult to predict solar power and wind power.

Introduction of renewable energy source cause fluctuations in the energy production. The fluctuations are usually managed via import and export of electricity to neighbouring countries and with the spinning reserve. However, the difficulties in handling the fluctuations increase as the share of renewables in the energy system is raised.

The share of electricity produced from wind turbines in Denmark today is close to 20% of the total electricity production (Danish Energy Agency 2009). With the current energy plans, this percentage will increase in the future (Danish government 2011). Today the production of electricity is adjusted to the demand for electricity. This demand fluctuates so that demand generally is low during nights,

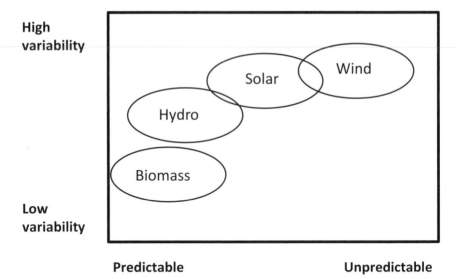

High
variability

Low
variability

Predictable   Unpredictable

Figure 2.1.4 Balancing renewable energy sources in the energy system

higher during the day and peaks when consumers return from work and switch on their lights and appliances. Electricity supplied by wind turbines furthermore makes the energy production fluctuate according to weather conditions which means that the energy system is challenged by fluctuations both in the demand and in the production.

Figure 2.1.5 illustrates an average day in January 2011 in Western Denmark, hour by hour. The purpose with the figure is to illustrate the structural relations between components in the energy system. Note that wind power production fluctuates on a day by day basis.

What is interesting about electric transport is that a systematic and intelligent coordination of battery charging makes it possible for the grid operator to absorb fluctuations in electricity production if batteries are charged while there is a surplus of energy on the grid. Such a system, usually referred to as a *smart grid system*, also has a benefit for the consumer as it makes it possible to charge batteries when electricity prices are low. Charging electric cars at night would furthermore enable utility companies to improve the capacity utilisation of existing power stations as vehicle charging at night would level out fluctuations in demand and enable utility companies to run a stable power supply.

Such a system however requires an intelligent battery charging system that would enable communication and coordination between components of the energy system and the individual battery charging stations (or the individual car if the technology is fitted on-board). This is on the one hand a challenge to the established energy and transport sector but on the other hand an opportunity to companies that can manage to provide the needed solutions.

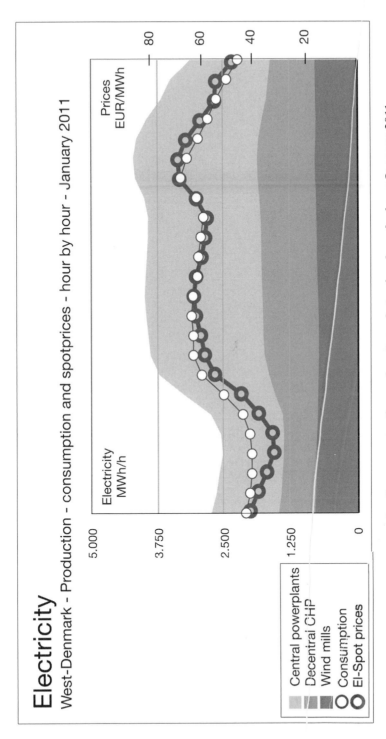

Figure 2.1.5 Average electricity production, consumption and spot market prices hour by hour January 2011

Exploiting the Opportunity of Electric Transport

The implementation of renewable energy sources in the energy sector therefore opens a window of opportunity for implementation of CleanTech solutions related to electric transport. A number of different business models can be derived from the opportunities illustrated above. Kempton and Tomic (2005) identify three different business models.

1. *The fleet operator model:* where owners of large numbers of electric vehicles (typically trucks, busses etc.) coordinate charging according to peaks in the electricity production. This is simple to implement but the overall impact on the energy system is fairly insignificant due to small number of fleet vehicles compared to private cars.
2. *The utility model:* a business model based on existing relationships between utility companies and consumers (in this case electricity consumers that also own electric cars) in which utilities manage and coordinate intelligent charging directly with the consumers. This business model holds a great potential but the utilities have for various reasons so far not shown interest in exploiting the opportunity. Exploiting the opportunity requires a rethinking of the current electricity market which acts as a barrier.
3. *The Third party model:* a business model where a third party act as a mediator between the energy market and electric car owners.

Kempton and Tomic (2005) originally developed the models for a more advanced vehicle-to-grid system where the electric cars also supply power back to the grid. Vehicle-to-grid services are not likely to be implemented in the near future in Denmark and have for that reason been ignored here.

The lack of interest from utility companies and car manufacturers illustrate an interesting point: Established companies tend to focus on core competences, which in the case of energy companies is production of electricity and in the case of car manufacturers is design and production of cars. None of them seems to target the new market opportunity directly even though some energy companies, like DONG Energy, holds a minority shares in a company designed on the basis of the third party model.

One company that seek to exploit the opportunity is called Better Place. Better Place aims at selling electric car transport solutions and act as a mediator between electric car owners and energy producers. In doing so the company have developed a business plan that include the installation of charging stations, the installation of automatic battery changing stations, development of technical infrastructure to control charging of batteries and communication with electric car owners and energy producers. The idea is that the car owner buys an electric car without a battery and then subscribe to a package that allows the car owner to use batteries and infrastructure installed and owned by Better Place. The driver pays Better

Place according to the millage he drives. This business model enables Better Place to earn money from controlling, coordinating and charging the batteries while prices on the energy markets are down due to low consumption or high wind power production.

Conclusion

The CleanTech concept is closely related to existing concepts such as cleaner production and eco-innovation however it also adds a new dimension by focusing on entrepreneurship, new business models and financing. The theoretical discussion presented three different types of CleanTech opportunities: 1) opportunities that can be identified and exploited within the boundaries of single companies, 2) opportunities that are best identified and utilized in collaboration with stakeholders associated with the lifecycle of a product or service, and 3) CleanTech opportunities that are best identified in a system perspective.

Exploiting CleanTech the opportunities often strand when a novel technology is invented but not yet commercially viable. This is referred to as the valley of death. Bridging across the valley of death often requires that the business model that is implemented is able to simultaneously provide solutions to multiple problems instead of solely relying on advantage of a single break-through technology. Focusing on providing solutions to multiple problems often boosts the economy as revenues are generated from various sources. Such a system is illustrated with case of electric transport.

References

Baas, L. 2005. Cleaner production and industrial ecology, Dynamic Aspects of the introduction and dissemination of new concepts in industrial practice, Eburon Academic Publishers, Delft, The Nederland's

Berkel, R. 2000. Cleaner production for Process industries, Overview over the cleaner production concept and relation with other environmental management strategies, Plenary lecture – CHEMECA 2000

Burtis, P.R. 2004. Creating the California CleanTech Cluster, How Innovation and Investment Can Promote Job Growth and a Healthy Environment, Natural resource defence council and environmental entrepreneurs, copy obtained from http://www.nrdc.org/air/energy/CleanTech/CleanTech.pdf, [accessed 9th of September 2011]

Brøndum and Fliess. 2009. *CleanTech – guldægget i dansk* økonomi, *Kortlægning af CleanTechfeltet i Danmark* , Copy obtained from http://www.cphCleanTech.com/media/59401/CleanTech%20guld%C3%A6gget%20i%20dansk%20%C3%B8konomi1810095.pdf, [accessed 9th of September 2011]

Cooke, P. 2008. CleanTech and an Analysis of the Platform Nature of Life Sciences: Further Reflections upon Platform Policies, *European Planning Studies* Vol. 16, No. 3, April 2008

COWI 2008, Promoting Innovative Business Models with Environmental Benefits, final report prepared for the EU Commission, DG Environment, November 2008, copy obtained from http://ec.europa.eu/environment/enveco/innovation_technology/pdf/nbm_report.pdf, [accessed 9th of September 2011]

Danish Energy Agency. 2009. Energy statistics, Danish Energy Agency, Copenhagen, Denmark

Danish Government 2011. Energistrategi 2050, published by the Danish Governemtn, February 2011, available at http://www.kemin.dk/Documents/Klima-%20og%20Energipolitik/Energistrategi%202050%20-%20final.pdf, [accessed 9th of September 2011]

Directive 2008/1/EC of the European parliament and of the council of 15 January 2008 concerning integrated pollution prevention and control, *Official Journal of the European Union* 29.1.2008

European Commission. 2004a. Facing the Callenge – The Lisbon Strategy for Growth and Employment, Report from the High Level Group chaired by Wim Kok, November 2004

European Commission. 2004b. Communication form the Commission to the Council and the European Parliament, Stimulating Technologies for Sustainable Development: An Environmental Technologies Action Plan for the European Union, COM (2004) 38, Brussels, 28 January 2004

European Commission. 2008. Communication form the Commission to the Council and the European Parliament on the Sustainable Consumption and Production and Sustainable Industrial Policy Action Plan, COM(2008) 397 final, Brussels, 16.7.2008

European Environmental Agency. 2009. *Transport at a crossroads, TERM 2008: indicators tracking transport and environment in the European Union*, EEA Report No 3/2009, European Environmental Agency, Copenhagen, Denmark

Geels., F.W. 2005. *Technological Transition and system innovation, A co-evolutionary and social-technical analysis*, Edward Elgar Publishing Limited, USA

International Energy Agency. 2011. Technology Roadmap – Electric and plug-in hybrid electric vehicles, Updated version June 2011, available at http://www.iea.org/papers/2011/EV_PHEV_Roadmap.pdf, [accessed 9th of September 2011]

Kemp, R.. 1997. *Environmental Policy and Technical Change – a Comparison of the Technological Impact of Policy Instruments*, Edward Elgar

Kempton, WW and Tomic. T 2005. Vehicle-to-grid power implementation: From stabilizing the grid to supporting large-scale renewable energy, *Journal of Power Sources*, Volume 144, Issue 1, 1 June 2005, P. 280–294

Kuehr, R.2007. Environmental technologies from misleading interpretations to an operational categorisation & definition, *Journal of Cleaner Production* 15 (2007) p. 1316–1320

Lundvall, B. Å., 1992, *National Systems of Innovation, Towards a Theory of Innovation and Interactive Learning*, Pinter Publishers, London

Mol, A.P.J, Jänicke, M. 2009. The origins and theoretical foundations of ecological modernization theory, in *The Ecological Modernisation Reader*, edited by Mol, A.P.J, Sonnenfeld, D.A., Spaargaren, G, Routhledge , Taylor and Francis Group, London and New York

Murphy, L., M. and Edwards, P., L. 2003. *Bridging the Valley of Death: Transitioning from Public to Private Sector Financing*, National Renewable Energy Laboratory, U.S. Department of Energy Laboratory

OECD. 2000. Policies, strategies and recommendations for promoting cleaner production in developing countries, OECD Working Party on Development Co-operation and Environment

Pernick R. and Wilder, C., 2006, The CleanTech revolution, The next big growth and investment opportunity, Haper Collins Publishers, New York

Pernick R, Wilder, C. And Winnie D., G., C., 2009, Clean Tech Job Opportunities 2009, Clean Edge

Porter, ME and van der Linde, C. 1995, Green and Competitive, Harvard Business Review, September-October 1995

Porter, M., E. 1998 *On Competition*, A Harvard Business Review Book, Harvard Business School Press, USA, Boston

PricewaterhouseCoopers, 2009, *Adopting electric vehicles: The role of technology and investment*, October 2009, available online at http://www.pwc.com/ar/es/publicaciones-por-industria/assets/adopting-electric-vehicles.pdf [accessed 9th of September 2011]

Thrane, M. and Remme, A. 2007.Cleaner Production, in *Tools for Sustainable Development*, edited by L. Kørnøv, M., Thrane, A. Remmen and H. Lund, Aalborg Universitetsforlag 2007, Aalborg Denmark

US EPA 1988, *Waste Minimization Opportunity Assessment Manual*, Hazardous Waste Engineering Research Laboratory Office of Research and Development US Environmental Protection Agency Performance, United States Environmental Protection Agency, USA

UNCED, 1992, Agenda 21: Chapter 34, *Transfer of environmentally sound technology, cooperation and capacity building*, UN Department of Economic and Social Affairs – Division for Sustainable Development, Earth Summit, Rio de Janeiro, Brazil, 1992

UN DESA 2007, *3rd International Expert Meeting on 10 Year Framework of Programmes on SCP (Marrakech Process), Background paper 2: Key Issues of Sustainable Consumption and Production, Intended to support discussions in Working Groups on 27 June 2007*, Stockholm, Sweden – 26–29 June 2007, available online at http://www.uneptie.org/scp/marrakech/consultations/

international/pdf/StockholmConferencePaper2.pdf [accessed 9th of September 2011]

UNEP. 1996. *Cleaner production in Wet Textile Processing – A workbook for Trainers*, United Nations Environment Programme, Industry and environment, First Edition, March 1996, available at http://www.unep.fr/shared/publications/pdf/WEBx0033xPA-TextileWet.pdf, [accessed 9th of September 2011]

UNEP. 1998. International Declaration on Cleaner Production

UNEP, 2000, *Cleaner Production Assessment in Dairy Processing*. Prepared by COWI Consulting Engineers and Planners AS, Denmark for United Nations Environment Programme; Division of Technology, Industry and Economic; April 2000, available online at http://www.unep.fr/shared/publications/pdf/2480-CpDairy.pdf, [accessed 9th of September 2011]

UNEP/SETAC 2009, *Life Cycle Management – How business uses it to decrease footprint, create opportunities and make value chains more sustainable*, United Nations Environment Programme, Division of Technology, Industry and Economics and Society of Environmental Toxicology and Chemistry, available online at http://www.unep.fr/scp/publications/details.asp?id=DTI/1208/PA [accessed at 9th of September 2011]

WBCSD 2000, Eco-efficiency – Creating more value with less impact, World Business Council for Sustainable Development, available at http://www.wbcsd.org/web/publications/eco_efficiency_creating_more_value.pdf [accessed 9th of September 2011]

Åhman, M., 2001, Primary energy efficiency of alternative powertrains in vehicles, *Energy,* 26 (2001) p. 973–989

Chapter 2.2

Local Transition Strategies for Low Carbon Construction and Housing – Studies of Innovative Danish Municipalities

Jesper Holm, Inger Stauning, and Bent Søndergaard

Introduction

Mitigation and adaptation to climate changes have become a main issue in policy making and planning in a range of sectors and on different policy levels. Whereas hesitance and symbolic policies are most often what we have witnessed, when it comes to national mitigation policies, this vacuity seems to have spurred local municipalities to develop different kinds of climate policies, action plans and projects that go beyond any mandatory requirements from national governments (Bulkeley and Kern, 2006, Späth and Rohracher 2010, Montin 2007, ICLEI 2010). Recently, worldwide, this effort has been formulated in voluntary agreements among regional authorities such as ICLEI's *Urban CO_2 Reduction Project* and *Cities for Climate Protection* campaigns, or *The World Mayors and Local Governments Climate Protection Agreement* from COP14 in Bali 2007, and EU's follow up: *Covenant of Mayors* from 2007.

Within this global regime of climate politics among local and regional authorities, the chapter discusses the type and characteristics of new forms of governance in the climate policies of Danish municipalities. The chapter builds on our studies of municipal climate adaptation and mitigation policies in Denmark, in these studies we have paid attention to efforts made at lowering energy consumption in construction – new dwellings, refurbishment and transformation of energy supply.

Energy consumption related to buildings accounts for 40% of the total Danish energy consumption – making it a focal part of national and local climate strategies and programs. For this reason, examining innovative local initiatives becomes highly relevant. In this chapter, *climate mitigation in the construction and housing sector is defined as the promotion of non-fossil energy and energy saving in the combined field of energy systems and construction/housing sector*. The development of local climate strategies and policies is analysed from two perspectives: the shift of governance forms and the transition of socio-technical systems:

- A new generation of regulatory efforts for municipalities with new forms of capacity building and governance will be required for the firm development

of low carbon societies. Developing climate strategies and action demands
multi-actor and multi-level/scalar governance.

- Local mitigation and transition to low carbon societies requires a transition
of socio-technical systems, thus local governance will have to build
capacities to influence and transform local socio-technical systems.

The two perspectives are highly interrelated – the character of the challenge
(transition of systems) rules out more traditional policy approaches, and the local
strategies and programs adopted have to be judged by their ability to provide
capacity building for the transition of systems. Climate strategies may be initiated
by various other means like campaigning and sector planning, but here we want to
examine how municipal policies and strategies address and influence actors, norms,
concepts and discourses around existing local or national socio-technical systems.
What are the transitional outcomes of climate strategies and policies? How do
local emission policies related to the combined field construction-energy supply
influence and shape the transitions of the socio-technical systems?

The development of climate strategies and programs in Danish municipalities is
an ongoing process. A number of municipalities have a long record of initiatives and
programs, whereas most of the Danish municipalities are in the process of defining
their climate strategies and programs. They are in a search and learning process.
The chapter analyses the strategies and processes of a few of the frontrunners and
discusses the role of municipalities in the development of low carbon societies,
focusing on the Danish conditions and transition of socio-technical systems.

The chapter builds on a three year R&D project, Climate Change and
Innovation in the Building Sector (CIBS 2011), established to examine how Danish
municipalities and regions may integrate climate mitigating, and sustainable
construction and refurbishment principles in regional business and municipal
strategies of housing and construction (http://www.climatebuildings.dk). On the
basis of multi sourced databases, expert interviews, field trips and seminars we
have established an overview of current construction/energy initiatives among
the municipalities (for an overview, see CIBS 2011). This has led to more in-
depth case studies of types of political-administrative initiatives, in order to
identify various strategic orientations, actor configurations, and which type of
socio-technical experimentation or planning efforts that have occurred. It is an
explorative study. The four municipalities which we discuss here were not chosen
for their representativeness, but for having adopted and implemented a variety of
new approaches to local climate governance in relation to the combined energy
and construction sectors. This incurs a bias; other municipalities could have been
selected with the effect that other issues and problems would have been exposed.

Studies of frontrunner municipalities which are undertaken so as to learn from
early movers about the feasibility, potentials and obstacles of new local climate
governance and its ability install local transition processes, demand qualitative
studies, and hence this study has a limited sample of four municipalities.

Institutional Capacity Building for Transition Politics

Planning and policy studies have raised the question of how we can understand the preconditions for the emergence of more sustainable transition processes and how we undertake such processes (Kemp and Rotmans 2001, Voß and Kemp 2006). A multi-level model of niche-regime-landscape has been suggested in which regime shift processes, such as establishing renewable energy systems or low or energy-plus housing, are understood in terms of an interplay between socio-technical regimes, socio-political landscapes and niches (Geels 2004, Kemp and Rotmans 2001). The current socio-technical systems of the combined energy-construction field can be characterised by dominant regimes, sustained and supported by predominant actors, institutions, standards and regulation, maintaining the technological trajectories. These include the dominant fossil energy regime, standardised refurbishment materials and methods, or the pre-fab, fordistic single house construction. Other socio-technical paradigms of more sustainable and climate adapted houses are emerging, but managing such shifts would take structural changes in terms of reconfigured regimes.

The dominant regime may be changed by innovative processes performed by networks of the regime actors, driven by dynamics on the different dimensions of the regime, e.g. industry, policy, science, markets. These processes may be influenced by changes in the landscape, such as rise in energy prices, shifting agendas or the creation of new national conditions for development, making "windows of opportunity" for new technology systems. Regime changes may also be influenced by niches of emerging new technologies and ideas and inspiration from experimentation among social entrepreneurs, alternative construction actors and so on. Processes of niche creation, learning processes, translation and transfer of ideas are very important for the transition, and innovative actors in as well as outside the dominant regime are here of interest (Geels 2004).

Thus, one of the instruments for sustainable transition can be niche management, i.e. creating and supporting niches for experimentation and maturation of new construction, energy and mobility technologies (Kemp, Schot and Hoogma 1998), so they can be adopted by innovative mainstream companies and organisations and disseminated in the dominant regime (Kemp and Loorbach 2006).

In such transition processes actors in local communities may play significant roles, especially in the void of national strategies. Room for experiments and local networking on new forms of energy supply, building and retrofitting of houses can be shaped and facilitated in the local area. An important issue here is how such experimentation and deliberation for new construction-energy utilities, standards etc. may assist in the shaping and configuration of new actors, networks, institutions and settings. A second issue is how they may serve as arenas for searching and learning in new energy technologies, new construction and refurbishment technologies or new building materials and processes. In this perspective municipalities hold the potential for becoming situated transition places (Holm, Søndergård and Stauning 2011).

Thus we are interested in how municipalities may assist in forming new actor-network configurations, and in establishing, managing and diffusing socio-technical experimentation. We presume that this requires the building of an institutional capacity among the municipalities` political and administrative sectors, for the engagement of market actors, social movements, house owners, the financial sectors etc.

What are the governance capacities means required for shaping distributed, common deliberation in climate adaptation, in public-private R&D efforts, in distributed planning for local energy and construction standards? To learn about this, we are interested in which type of institutional arrangements and which strategies and capacity building exist among frontrunner municipalities. To this end we have studied concepts and experiences from promising, climate related, energy and construction efforts in selected Danish municipalities.

Danish Government Climate Policies Addressing Local and Regional Authorities

Danish municipalities enjoy relative autonomy on local matters. They are subject to financial steering and regulation from central government in terms of compulsory orders defining certain obligations (e.g. social care, education of children, spatial planning, waste handling, public service activities, energy provision, and environmental protection), but in many fields they are given discretionary power allowing for flexibility in fulfilling local ends.

Decisions regarding construction and the renovation of buildings, public transport and spatial and sector planning for energy and construction, are (following national statutory orders) partly made locally and have the potential to either promote or disregard climate friendly solutions. A structural reform in 2006 (*kommunalreformen*) created bigger and stronger municipalities which were given extended responsibilities in planning and regulation, nevertheless national policies and planning for energy systems and for construction standards predominate.

Central government has addressed climate problems within a cost-benefit approach, embedded in a neo-liberal perspective. The "Climate strategy" (Regeringen, 2003) focused on how market forces could be used to meet the Danish CO_2 reduction obligations in an economic way, paying low attention to local development of low carbon societies. The program 'The Business Strategy on Climate Change' (Regeringen, 2009) represented a major turn, but was not on development of low carbon societies. Main issues in the strategy were to address climate as a driver of business and innovation and examining how climate (cleantech) clusters and partnerships could be identified and developed. Neither of the programs had a transition perspective or identified new roles for municipalities. The strategy paper also represents a lack of a coherent climate strategy, as only ad hoc adaptation efforts are sought.

The need for a coherent climate strategy has to some extent been acknowledged by the Danish Government. March 2008, the government appointed a "Climate commission" (Regeringen, 2008) mandated to develop 'energy and climate policy instruments' enabling a phase out of fossil fuels in the Danish energy system. Their final report (Klimakommissionen 2010) identified a fossil-free Danish energy system in 2050 as a realistic goal, and forwarded a recommendation for a number of policy instruments to reach that goal. These recommendations, so far, have not been turned into new initiatives, but are (June 2011) subject to negotiation in relation to a new national agreement on energy.

Thus, we have a global agenda, but lack of effective international and national action. We have a need for developing low carbon communities, but there is no national support or coherent strategies for how to start this transition. Despite growing concern on climate and energy issues at international and national levels, central government has not defined a clear role of Danish municipalities in the development of low carbon societies. Municipalities have been given some specific obligations in relation to climate and energy issues, as using mandatory energy assessment schemes for municipal buildings, but on a general level no strategic visionary aims and no statutory duties have been issued by central government. This lack of centrally defined framework leaves the municipality in a situation where they will have to invent new policies and instruments, if for some reason there are local drivers and motivations for innovative climate politics.

Climate Strategies in Municipalities – Overview and Characteristics

Most of the Danish municipalities are in the process of developing energy and climate policies and plans. Sperling, Hvelplund and Mathiesen (2010) found that 63 out of 93 responding municipalities (early 2008) were working with climate issues, and a survey media 2010 (Mandagmorgen 13.8.2010) reported that 91 out of 98 municipalities were working actively with climate issues and had plans for GHG-emission-reduction. It is part of the tendency that the municipalities subscribe to what can be labelled semi-private political arrangements and networks of voluntary action. Sixty-nine of the 98 Danish municipalities have signed a commitment to be a member of the 'Climate Communities Club' ((*klimakommuner*), a framework institutionalised by the Danish society of Nature Conversation, promising an annual reduction of CO_2 on 2 % or more, www.DN.dk/klimakommuner (1.12.2010). And 21 Danish municipalities have joined the EU 'Covenant of Mayors' programme as a framework of voluntary local climate action.

In our more in-depth study of the region of Zealand, we found that 12 out of 17 municipalities are members of the 'Climate Communities Club', and 14 have signed the EU voluntary commitment scheme Covenant of Mayors. More than half of them are in an implementation process of elaborating climate plans with the objective of developing a platform for climate action. It implies mapping of local CO_2 emission identification of reduction potentials, and, further on, a definition of political goal and strategies, including building of local capacity. But it also implied experimenting with new concepts of governance to include local actors and citizens. (Stauning, Holm and Søndergård 2011).

The roads of the municipalities have varied. In most cases the development of basic data, plans and strategies has been quite rationalistic based on statistical mapping of emissions and identification of strategic potentials for CO_2 reduction. Often consultants have undertaken the development of plans and strategies, only few municipalities have had an internal elaboration of plans and/or have adopted more bottom-up participatory processes involving local actors and citizens in identifying goals and actions. The Covenant of Mayors, now signed by most of the municipalities in the Region of Zealand, commits the municipality to go beyond EU norms for energy action plans, and networking with other cities. This provides an opportunity for making a shared framework enabling a collective development of knowledge and competences among the municipalities in the region.

Climate policies and strategies have become part of local policy in municipalities. Still, however, the development of local climate strategies in Danish municipalities is at an initial stage, both in terms of the development of local institutional and strategic capacity for the process, and in terms of obtaining the right conditions by central government. To investigate the potentials and look closer at the strategies and capacity building, we have looked at how municipalities have addressed energy reduction in construction and housing focusing on how they targeted the socio-technical systems by strategies to incorporate and mobilize their actors, knowledge and discourses and (private businesses and citizens).

Municipalities as Arenas for Emerging Climate Strategies – Energy and Housing

The reduction of energy consumption in housing in Denmark has a long history. The energy crisis in the early 70s gave way for a first wave of energy refurbishment of the housing stock and introduced an energy conscious technology path in the construction sector as well as in the energy sector (e.g. rolling out district heating based on combined heating and power plants). On an institutional level, a tax on energy and CO_2 was introduced, subvention schemes for insulation were established, new construction standards for insulation were enacted, and covenants for corporate energy savings were issued. Housing communities and green movements pushed for and experimented with even more radical changes, demanding sustainable construction and the integration of renewable energy solutions (Holm, Stauning and Søndergård 2009, Gram-Hanssen and Jensen 2004).

As the climate agenda at the turn of the century redefined the long term goals of housing and construction, the Danish situation was characterised by:

- A building stock showing very high differences in energy standards and still holding a high potential for energy saving

- A construction sector, characterised by traditional technology, low efficiency, low innovativeness and low environmental awareness
- Availabilities of new energy efficient technologies, demonstrated in local projects or abroad, but not integrated in the construction sector.

In the last decade we have witnessed a number of initiatives from central government (see table 2.2.1) and sector actors, but the transition to a higher level of energy technology in construction has been slow. Experimentation and innovative set up's have to a high extent been driven by individual actors and local initiatives, and, as we shall see, typically have involved active municipalities (Holm, Stauning and Søndergård 2009).

Table 2.2.1   Central government initiatives for enhanced energy efficiency in buildings

Central government initiatives (selected):
<ul><li>Redefined construction codes, gradual new buildings</li><li>Energy strengthening of energy efficiency of standard label of buildings. Mandatory in relation to sale. Made mandatory to municipalities</li><li>Permission to integrate energy efficiency demands in local plans (municipalities)</li><li>Mandatory energy reduction demands to energy companies</li></ul>

Danish Municipalities as Actors in the Energy and Building Sectors

The energy and housing/construction system involves a broad array of actors ranging from individual households, real estate/property companies, district heating societies, private enterprises in construction, local distribution companies to international energy companies. It is subject to a broad range of regulations and both national and local institutional settings.

Municipalities are both involved as operator and regulator of the energy system. Historically, local communities and municipalities in Denmark have played a major role in the development and operation of local energy systems; electricity, gas and/or heating has been supplied by public utilities (often run by municipalities) or by user owned societies (often incorporating municipality representatives). Still, in the new structure of the energy sector (following the EU's energy market approach) municipalities and user-owned energy companies constitute major elements in many local energy systems. As operators, municipalities have to operate public utilities on a non-profit basis and without cross subsidisation, however, they are allowed to include fulfilment of environmental and climate objectives in their decisions and projects.

As regulators, the municipalities hold the responsibility of heating planning (local discretionary power, however, is subject to the confines laid down in national heating planning reserving zones for district heating and the natural gas grid). Along national regulation municipalities have been involved in establishing district heating companies for mainly urban areas, so the CO_2 efficient district heating system today covers 60 % of Danish homes. Municipalities also have participated in promoting local combined heat and power plants, waste incineration and biogas power plants, and in their capacity as spatial planning authorities, municipalities also hold responsibility for locating power plants and district heating plants. However, how municipalities have utilised their roles has varied, and across the municipalities, we have a scattered picture, there is a great deal of variety in the local structure of the energy sector and their interconnection with the housing sector.

Municipalities can directly influence construction related CO_2 emissions in their operation and management of energy consumption (use of electricity and heating) within public owned buildings, institutions etc. Through local policies on building standards relating to new public construction, energy management and refurbishment, they both can curb emissions and serve as role models. Green purchasing and environmental management schemes has here played a role for a more conscious green management culture among municipalities.

When it comes to the private housing sector and energy-climate aspects, municipalities have the overall planning mandate for spatial planning at municipal and local level. It gives them authority to decide on 'energy profiles' of new settlements/district areas, e.g. decisions on dispersed single family homes or condensed cities with apartments, on car-parking space at dwellings or not, on the type of heating system. Besides, local authorities are mandated to control and guide private constructors in the building of houses. Technical standard, including energy standards are issued in the national building regulations, but municipalities can take a role in terms of developing local skills, traditions and cultures and by utilising their mandate to enforce the use of high energy standards in new buildings.

Thus, the municipalities have many different roles, but at the same time they are not primary actors: the decisions of how to build and which materials and energy systems to use are in the hands of the private building owners, the building companies and the energy suppliers. In order to influence the sector, they have to use governance forms that motivate and influence the private actors to change their technology. Here Danish municipalities have a long tradition of private-public cooperation and consensus-seeking approach, involving different local actors; a tradition for participatory and mobilising activities which had been further developed in Local Agenda 21-initiatives.

Policy Culture of Deliberation

The emergence of a deliberative policy culture for climate initiatives among the Danish municipalities can be related to the decentralised public administration, a consensus-seeking and entrepreneurial approach in policy style, and a tradition of 'popular enlightenment' (Læssøe 2000). These factors provided a favourable landscape for an influential implementation of Local Agenda 21 in Denmark (Holm 2010). Thus, experiences with Local Agenda 21 entrepreneurial cultures, strategic co-operative management regimes, and niche experimentation on energy-, organic food, eco-housing, urban ecology etc. have paved the way for current climate mitigation initiatives among the frontrunner municipalities we have investigated.

A tradition characterised by public participation in local planning and reflexive regulation of environmental control and standards for companies, has also prepared a political-administrative regime among local authorities for addressing socio-technological system actors. In addition, the multi-partisan and multi-level tradition (Jänicke 2000) that incorporates plural interest groups in the design and implementation of local policies together with a comprehensive number of local, green 'do-it-yourself' experiments has made it relatively easier for officials to initiate LA21 and currently instigate climate policy projects with a considerable degree of public interest.

Of most importance for this chapter is that the local Agenda 21 experiments have revealed a new paradigmatic way for developing a separate path in municipal policy, where political mobilisation for sector crossing environment and business initiatives forms new *visions for local development, community cohesion and social welfare*. New partners have been found for a number of environmental areas that were not under the rules of hitherto environmental acts and regulations. They formed the basis for a change process – including citizens, NGOs, business and authorities -in more comprehensive and constructive efforts to re-build cities and infrastructures, to engage health and social care activities in greening. Resource accounting, quality of city-life and environmental goods became a positive focus in an optional oriented policy feature, instead of protecting the environment through restrictions on activities, the 'bads'.

Governing LA21 has, beside a lot of window dressing and simulation politics (Blüdhorn 2005), been a vehicle for new types of governance beyond the scope of the *rechtstaat or the welfare state;* LA21 policies in Denmark have especially among front runner municipalities fostered i.e. imaginative community policies, experimentation, partnership politics, and corporate and civic deliberation. Accordingly Danish LA21 policies have fostered sector crossing initiatives and public-private partnerships with *niche* efforts of creating protected spaces of alternative energy supply, resource saving; test bed for urban ecology show cases. Among these efforts, also environmental and energy infrastructure coupled to dwellings has been an issue.

Frontrunner Municipalities – Demonstrating the Range of Strategies and Means

Many municipalities have taken up the policy culture of deliberation, entrepreneurship and new governance forms of public-private partnerships, and brought them into local climate policies. Individual municipalities have adopted high profiled goals and strategies for development of low carbon local societies (e.g. Frederikshavn, Samsø, Sønderborg, Lolland, Egedal) and in parallel municipalities have engaged in national trans-municipality co-operations (Dogme 2000 (now Green Cities), Bycirklen) as well as international cooperation (e.g. Aalborg Commitment, ICLEI (International Council for Local Environmental Initiatives, Covenant of Mayors). Within the framework of Local Government Denmark (KL Kommunernes Landsforening) environmental planners (and the likes) have formed a professional network (practice community) on municipality planning on environment, energy and climate.

The next section presents constitutive elements of climate policies in some of the frontrunner municipalities to characterise the development of instruments and governance forms, showing the scope of approaches and experimentation/ experiences when it comes to transition of socio-technical systems of construction-energy.

Sønderborg

In addition to many years of Local Agenda21-activities, the Municipality of Sønderborg has, together with local stakeholders set the goal of being CO_2 neutral in 2029, and together with local actors they have organised a public-private partnership 'ProjectZero' (2007) to fulfil this goal. Local industrial actors (e.g. Danfoss), energy companies and institutions/organisations have engaged in the partnership as a strategic project, identifying the local transition to a low carbon society as an important platform for socio-technical learning, innovation and business development. For the municipality, the vision is that the project of CO_2 neutrality can be a driver of local development.

ProjectZero can be characterised as a hybrid business-climate mitigation organisation involving and building on the resources of local stakeholders. It serves as an arena of local dialogue and thus contributes to the building of shared local visions. The hybrid organisation, operating with independent funding, has paved the way for initiating and engaging in development and experimentation projects which otherwise would have been difficult to undertake by municipality organisations (due to restrictions laid on how municipalities may engage in projects involving business). ProjectZero has established a strong organisation, and performed a local capacity building. This has included cooperation with knowledge institutions. An outcome of this has been the development of a roadmap for local CO_2 reduction 2010–15 (Sønderborg 2009). For this work, the management scheme and ability to define projects and processes, ProjectZero 2010 was awarded the European Commission's prestigious award for the best sustainable energy solution (category "Sustainable Energy Communities").

The ProjectZero roadmap defines a holistic approach. Taking abatement of housing related CO_2-emissions as an example, reduction is to be obtained by a concurrent effort on the substitution of fossil fuel, improved energy efficiency of energy infrastructure and – provision, energy refurbishment of the building stock and changed behaviour. In parallel, the municipality enforces low-energy standards for new buildings, also aiming to spur the development of the local construction industry. The program outlined in the roadmap, backed by the cooperation of actors in ProjectZero, serves to emphasise the need of concerted action, and stipulates roles and obligations/deliveries of local actors of the energy-housing system – a first step in establishing a reconfigured local socio-technical system.

The project ZeroBolig [ZeroHousing], launched by ProjectZero to spur the energy refurbishment of private houses, adopts a multi actor approach, and deliberately seeks to reconfigure the local socio-technical system of housings and energy. It combines local campaigns setting an energy refurbishment agenda, the restructuring of access to energy advice, the instigation of an upgrading of competences of local craftsmen, the involvement of banks and energy providers and the undertaking undertakes of experimental projects (e.g. intelligent heat pump set up) and local demonstration projects (exemplars).

Defining private-public (hybrid) organisations and arenas to build local strategies and projects have been part of the development of energy and climate policies in many municipalities (e.g. Herning, Albertslund). The unique feature of 'Project Zero' has been the strong bond to and economic backing of local industrial actors, the radical vision, the high level of organisation and capacity-building. On the other hand, the municipality´s quite weak role in the co-operative management regime may be criticised for leaving it too far for local business to set the agenda. Still, however, the municipalities are subject to economic limitations curbing both the building of adequate capacities in the municipality and limiting its ability to support projects and experiments.

Kolding

The municipality of Kolding (http://www.climatebuildings.dk/kolding.php) took an early lead in the energy management of its own building stock and in the development of a local climate and energy strategy. By establishing management procedures and organisational capacity they have integrated a systematic effort to obtain local energy savings in public buildings by installing low energy consumer goods, integrating energy savings in refurbishing, developing user-adapted heating and lighting systems etc. In construction, Kolding was an early adopter of a general demand that all new constructions should meet the highest energy standards (Energiklasse 1).

The work has been rooted in an early adoption of the Aalborg Commitment (1994), participation in Dogme 2000 network (now Green Cities), and in the participation in ICLEI (International Council for Local Environmental Initiatives) – all commitments binding the municipality to define sustainability goals, monitor and report progress, and to root its work in local actors. More specifically, the municipality in cooperation with 'Business Kolding' has issued an energy plan, "Energy Kolding", setting the goal of reducing CO_2 emissions by 75% per capita in 2021 (compared to 1990=100, Kyoto Protocol reference year) and it has established an independent environmental department (climate and sustainability) under the technical department. In 2008 they were appointed as an 'energy municipality' within the campaign of the Ministry of energy and climate.

Kolding represents a long term building of local capacity in the internal organisation and in partnership relations, as competences and institutional capacity has been acquired in a long process starting with the local and inter-municipality Agenda 21 work. Kolding also hosted the early experimentation in urban ecology projects from the 1980's to 1990's, and they have several show cases of sustainable constructions' refurbishment.

Middelfart

Middelfart (http://www.climatebuildings.dk/middelfart.php) has become a frontrunner in examining how private-public partnership based on Energy Service Companies (ESCO) can be part of local energy and climate work. In cooperation with an ESCO-company, the municipality defined investments ensuring a 20 % reduction (operating with return of investment in 10 year) in energy consumption in the building stock of the municipality. The experiences in Middelfart (and from other ESCO-frontrunner municipalities) are used in a research project to develop concepts, roles and institutional framework for such Private-Public partnership (http://www.ebst.dk/escommuner). The municipality has extended the ESCO-approach to private residential homes, facilitating (as an experimental project) a packet solution of energy refurbishing to a local residential area. The packets involved a concerted action of energy companies (offering consultancy), financial institutions and construction firms offering a substantial discount if sufficient home owners took advantage of their offer.

Middelfart municipality represents a redefined division of work and new interface with the market and in this respect a new path of capacity building in municipality organisations and consultant/construction sector. It is a specific capacity to define and manage ESCO-projects, both to be acquired by the municipalities and process consultants.

Egedal

In the municipality of Egedal (http://www.climatebuildings.dk/egedal-municipality-info.php), many years searching for local Agenda 21 tools and strategies able to deliver more substantial achievements led to the development of a catalogue of local municipal strategies for enhancing more sustainable construction and energy use. In the first place, this was manifested in attempts to provide guidance on environmental issues in reviews of building licenses and in the operation of municipal buildings. This has facilitated the municipality to develop the necessary competences in environmental and energy technical issues, which at the time was only partly present in the construction industry.

By establishing an innovative way for enforcing mandatory low energy requirements for new dwellings, on sites owned by the municipality, Egedal became a frontrunner in showing the window of opportunities for new socio-technical systems of construction. By lobbying the central administration of the Danish government and by publishing reports on the issue, Egedal made it possible for all municipalities to have the option of drafting district plans, which operated with above-the-standard, mandatory energy class demands to houses built in the district. These efforts resulted in planning for a local sustainable construction housing area and the enforcement of guidelines for all new constructions to fulfil stricter energy class demands. These demands were manifested in the 2005 local Agenda-21 strategy and in an integrated sustainable construction strategy in the municipality spatial district plan. This was regarded an experiment concerning how to plan and build new housing districts of the community. Specific plans for nature, waste water, energy supplies and street lamps (diode light) were integrated into a large construction area for over 800 dwellings, with a set of specific demands on the constructors and owners of the buildings. Furthermore the planning of Denmark's largest thermal heating system is under construction for supplying the area.

An immense search process of more sustainable technologies and constructions had taken place by staff members in the municipality, as explorative studies of how to enforce the green construction ideas into juridically binding contracts etc. The actual development of the areas also required deep technical discussions with constructors and entrepreneurs, encouraging companies to look for the alternative technical solutions that were at stake. There was an underlying wish to influence the building sector and the market, and especially to make a show case for normal families of the potentials in normal, but sustainable dwellings.

Capacity Building and Strategic Bridging of the Energy and Construction Sectors

The cases provide evidence of municipalities taking different approaches in their climate and energy work. The sample of cases documents, however, that involving external actors, the development of hybrid forms of organisations and building of new capacities and competence in the organisation have been essential elements in implementing local, municipal climate policies. In this respect, local climate and energy policies are in line with local policy schemes in relation to Agenda 21 work and other new policy areas, such as health promotion strategies.

The development of climate and energy programs/plans takes a development of new governance forms, where both the development of new capacities and the involvement and building of relationship to the actors of the energy and construction sector are important elements. Part of the municipalities programs are deliberate strategic efforts to shape new local configurations of actors (new forms of local business techno-systems) capable of carrying through local transition processes for low carbon communities. In this transition perspective, the different strategies have to be judged on how they shape new configurations of actors, institutions and technologies in the local socio-technical systems.

The cases show new ways of involving the energy and construction sector: as partners in networks, as co-operators, as dialogue partners in communal (joint) development plans, but also the process of creating shared visions of future development and requirements, and in this way showing directions for innovations and planning, both in companies and among other stakeholders.

The next section examines in greater detail the range of governance instruments being available and being exploited by the Danish municipalities in relation to obtaining energy saving and spur transition to alternative energies in relation to housing and construction. The objective is to develop a more systematic account of the kind of governance undertaken by municipalities in this field. In this examination we address a) roles and instruments of the municipality in relation to housing and construction, b) governance forms and capacity building of the municipalities, and c) how local energy-housing systems are reconfigured among actors, institutions and technologies.

Findings on Climate and Energy Related Programs Among Danish Municipalities in Housing and Construction

a. Roles and instruments

We have argued that local climate policy, including energy saving and implementation of alternative energy, has become a main local policy field, and argued that this places an important obligation on and provides a challenge to local government. Municipalities are born the main actors in climate policies and energy/ CO_2 reduction in construction and housing, being the public authority operating at the junction of specific local conditions and national goals and policies, and being the public authority having the direct contact with the citizens. The 'ad hoc approach' of the Danish Climate strategy leaves the municipalities with an even higher obligation to take responsibility and be innovative, if firm climate mitigation is to become a political and business reality. The complexity of the challenge can be demonstrated by giving an account of the numerous roles municipalities have and can relate to in developing strategies and plans for energy saving in housing and construction (see Table 2.2.2).

Table 2.2.2   Optional roles of municipalities in relation to transforming construction and housing into low carbon units

- Agenda setter, deliberative envisioner
- Climate, heat & energy planner
- Networker among municipalities
- Manager of institutions and buildings (operation, refurbishment, new buildings)
- Spatial planner
- Landowner
- Entrepreneur, corporate networker
- Campaigner, framer, brander
- Provider of energy, heat and cooling
- Authority for constructing approvals
- Local business developer
- Purchaser
- Alter of hierarchical rule games
- Partner in international hybrid networks

In their searching and experimentation, municipalities have worked with climate adaptation and mitigation from different positions. Many municipalities have built on their positions as local planners and have developed local climate plans based on systematic sector mapping of local emission of CO_2 (and climate gasses), identification of reduction potentials and attempts to identify adequate instruments and programs. Others have developed their energy management capacity (e.g. Kolding) or have used their role as spatial/urban planner and landowner (e.g. Egedal) to develop local plans ensuring low-energy houses. In most cases municipalities seek to combine different roles/instruments, imitating, experimenting and learning from 'best practice'.

b. Governance forms

Numerous nomenclatures are in use for classifying local environmental policies, often dependent upon the subjects studied and the theoretical perspectives used. Thus, Bulkeley and Kern (2006) made a categorization of local climate polices that distinguishes between self governing (e.g. greening the municipality, inter-sector coordination), governing by authority (e.g. energy planning, mandatory energy standards for dwellings), governing by provision (e.g. renewable energy service systems), and governing by enabling (e.g. campaigns and loans). Similarly EU´s identifies four roles of municipalities within the Covenant of Mayors Programme:

The Municipality as a 1.Consumer and service provider. 2. Planner, developer and regulator. 3. Advisor, motivator and a role-model 4. Producer and supplier.

In both cases, a concurrent action within all dimensions is anticipated. Bulkeley and Kern (2006) make the explicit argument that we should operate with multiple modes of governing, and for this reason claim that great efforts have to be devoted to overcoming institutional ambiguity (related to the many agendas and interest related to the field of climate policy) and co-ordination. Although they assert all four roles of municipalities as important, they find that their studies indicate that '*modes of governing based on enabling other actors are coming to the fore in local climate protection climate policy*' (Bulkeley and Kern 2006:2251).

We are in line with this understanding. But based on studies, we believe that this finding should be taken one step further. Municipalities are not just enablers, but are strategic actors in the shaping of local configurations of networks of actors and technologies. We find that the cases presented give evidence of the central role of municipalities in shaping local socio-technical networks, including the shaping of shared visions and goals of such networks. If a wider set of material were included in the chapter – as the developments of the CO_2 neutral Island community of Samsø, the use of climate camps in the municipality of Herning, and the deliberate shaping of a triple helix cooperation in the municipalities of Albertslund (Gate 21) and Lolland (Test Facility Community) (http://www.climatebuildings.dk). – we would be able to provide a further indication that such roles of *strategic actor* in shaping local networks are central to local climate mitigation and adaptation policies.

c. New governance in a transition perspective

Climate and energy efforts and the new governance forms of municipalities developed in the energy – housing field have to be seen as part of a transitional process, where local socio-technical systems are reconfigured; new combinations of actors, institutions and technologies are developed as an effect of specific local projects or strategies/programs. In this perspective, the development of new governance does not only concern the capacity building of the municipalities as political and administrative entities, but also concerns the development of capacities in the local energy and housing systems as such. This type of capacity building is about bringing local actors and stakeholders together in networks structured by new agendas and new technology platforms.

Taking this extended perspective on municipalities as experimental arenas of new governance forms in relation to climate and energy, and on the capacity building, four major modes of governance for public-private interface on socio-technical systems can be outlined:

1. Building local competences and institutional capacity in the municipality: Building up knowledge and organisational space of manoeuvre for setting strict eco- and energy requirements in rules, tenders, plans and technical performance demands on: Construction, energy supply, materials, insulation, energy use, water supply etc. This is done in order to put a pressure on the mainstream development of business and technology and facilitate experimentation.

2. Networking for market based experimentation: Building up negotiation capabilities for flexibility and governance in dialogue with large companies or housing co-operations. Companies, cooperative housing societies and public building owners may through this networking urge the creation niches for socio-technical experiments. The municipal administration may give regulatory leeway to unfold the experimentation.

3. Enabling corporations and finances to create a new market on well-known technologies: Developing focussed campaigns and services on e.g. refurbishment in cooperation with construction businesses, and through dialogues with finance. This may be urged by CO_2 emissions mappings, climate surveys, the purchase of energy saving services and equipment etc. from consultancies and industries (E.g. ESCO concepts).

4. Participatory strategies for the shaping of shared visions: Discursive strategies using positive storylines about a new envisioned community, where all sectors contribute to creating a more harmonious nature-energy-community balance. A lot of effort is used in creating voluntary climate policy groups, and incorporating business communities in deliberation. The strategies also cover public-private initiated CO_2 management regimes among companies the making of a certain local climate brands of the municipality as a pioneers in renewable energy, energy efficiency, or in energy savings in institutions, construction etc..

Acting from these positions, municipalities as strategic actors co-shape local (and national) socio-technical networks. They are not to be seen as mutual exclusive forms, they can co-exist and often they need to be combined: networking for market based experimentation would also take some efforts in shaping a shared vision of the included stakeholders.

However, we have differences in how individual municipalities balance the different modes of governance. The presented cases may illustrate both these differences and the different outcomes of local strategies:

- The ProjectZero in Sønderborg combines an effort on shaping a shared vision and forming strong local networks and projects for market-based experimentation. They establish a strong business case on climate and energy

- The Middelfart ESCO projects combine an ambitious goal of energy saving with a use of upcoming market actors engaged in energy services. The outcome of this approach (followed by other Danish municipalities) is a configuration of ESCO-providers and a specific institutional capacity in terms of competences in consulting firms and municipalities, to manage market based energy-saving projects. The energy management in the municipality of Kolding, in this relation, gave impetus to competence configuration based on in-house capacity and a market for local enterprises

based on well-known technologies.
- The proactive use of local planning in the municipality of Egedal combined the development of institutional capacity in the municipality with a strong interaction with market actors in the construction industry. In short the outcome was an upgrading of both the main construction firms and developers and the formation of an exemplar defining the scope of local planning. The interactive learning among the actors was continued in the subsequent stages of the urban development.

Climate change mitigation and adaptation require radical transition strategies and innovation support policies to change production and consumption patterns in all sectors. Denmark, like the majority of national states, is in a condition of paralysis when it comes to introducing the necessary, radical climate policies. However, a number of regional and local authorities show promising creativity and new modes of governing in their capacity building and supporting of experiments in socio-technical systems, to track new pathways of low carbon economy and climate adaptation. Our study of innovative local practices among Danish municipalities, when it comes to combined energy and clean-tech related CO_2 reductions from the housing sector, has shown remarkable experiments in establishing new socio-technical niches, preparing for up-coming new socio-technical regimes and in planning for renewable energy based heating provision systems for the housing sector. The ability of municipalities' climate and energy policies to deliver energy saving and transition to renewable energy is important, but equally important is that these efforts will need to be amplified politically and economically by national polices if to succeed in enabling the formation of stabilised, sustainable socio-technical regimes.

Concluding Remarks

A number of regional and local authorities show promising creativity and new modes of governing in their efforts to build capacity, support of experiments in socio-technical systems, and to track new pathways of low carbon economy and climate mitigation. Our study of innovative local practices among Danish municipalities, when it comes to combined energy and clean-tech related CO_2 reductions from the housing sector, has shown remarkable experiments in establishing new socio-technical niches, preparing for up-coming new socio-technical regimes and in planning for renewable energy based heating provision systems for the housing sector.

Development of local climate efforts is still in an early stage. There is a need of further studies in local climate governance forms, and in particular how municipalities may develop governance forms capable to shape reconfigure local socio-technical systems. And in this relation address how local climate efforts and experimentation can be integrated and utilised in a wider transition process of socio-technical systems, such as the combined energy and housing system.

The ability of municipalities' climate and energy policies to deliver energy saving and transition to renewable energy is important, but equally important is that these efforts will need to be amplified politically and economically by national polices if to succeed in enabling the formation of stabilised, sustainable socio-technical regimes. Future studies should address how policies could be designed to facilitate local climate efforts in municipalities.

References

Blühdorn, I. 2005. Social movements and political performance. Niklas Luhmann, Jean Baudrillard and the politics of simulation, in *Macht – Performanz, Performativität, Polittheater*, edited by B. Haas, Würzburg: Königshausen and Neumann, 19–40.

Bulkeley, H. and Kern, K. 2006. Local government and the governing of climate change in Germany and the UK, *Urban Studies,* 43(12), 2237–2259.

CIBS, *Climate Change and Innovation in the Building Sector*. 2011. http://www. climatebuildings.dk (accessed 7.6. 2011).

Forsberg, B. 2002. *Lokal Agenda 21 för hållbar utveckling – en studie av miljöfrågan i tillväxtsamhället,* [Local Agenda 21 for sustainable development: A study of the environmental issue in a society of growth] (In Swedish with English summary), PhD Thesis, Umeå University.

Geels, F.W. 2004. From sectoral systems of innovation to socio-technical systems: Insights about dynamics and change from sociology and institutional theory, *Research Policy,* 33(6–7), 897–920.

Gram-Hanssen, K. and Jensen, J.O. 2004. Green buildings in Denmark. From radical ecology to consumer oriented market approaches, in *Sustainable Architecture*, edited by S. Guy and S. Moore, London: Marcel Dekker, 165–185.

Holm, J. 2010. Local experimentation and deliberation for sustainable development– Local Agenda 21 governance, in *A New Agenda for Sustainability*, edited by M. Figueroa et al. Aldershot: Ashgate,

Holm, J., Stauning, I., and Søndergård, B. 2009. *Climate and eco-adaptation in housing and construction – regional transition strategies,* 1st European Conference on Sustainability Transitions, 4–5 June 2009. Amsterdam

ICLEI 'International Council for Local Environmental Initiatives. 2010. *Local Governments for Sustainability*, http://www.iclei.org/index.php?id=global-programs (accessed 7.6.2010)

Jänicke, M. 2000. *Ecological Modernisation – Innovation and Diffusion of Policy and technology*, FFU Report 00–08. Forschungsstelle für Umweltpolitik. Berlin: Freie Universität Berlin

Kemp, R., Schot, J.P. and Hoogma, H. 1998. Regime shifts to Sustainability through processes of niche formation: The approach of strategic niche management, *Technology Analysis and Strategic Management,* 10(2), 175–196.

Kemp, R. and Loorbach, D. 2006. Transition management: a reflexive governance approach, in *Reflexive Governance for Sustainable Development,* edited by J-P Voß et al. Chelterham: Edward Elgar, 103–130.

Kemp, R. and Rotmans, J. 2001. The management of the co-evolution of technical, environmental and social systems, *International Conference towards Environmental Innovation Systems,* Garmisch-Partenkirchen, September 2001.

Klimakommissionen [Climate Commission). 2010. Grøn energi. (Green Energy). Final report.

Montin, S. 2007. Kommunerna och klimatpolitiken – ett exempel på tredje generationens politikområden [Municipalities and the climate policy – an example of a third generation policy field], *Statsvetenskaplig Tidskrift,* 109(1), 37–57.

MandagMorgen. 2010, Den kommunale klimarevolution [The municipal climate revolution], nr. 27, 13 August 2010, 17–22

Læssøe, J. 2000. Folkelig deltagelse i bæredygtig udvikling, in *Dansk naturpolitik i bæredygtighedens perspektiv,* København: Naturrådets temarapport nr. 2.

Pedersen, M. B. (2010): *Kommuner i Region Sjælland og deres klimaindsats ifht. byggeri og energirenovering* [Municipalities in Region Zealand and their climate efforts in relation to construction and energy refurbishing], CIBS Project working paper, RUC.

Regeringen [The Danish Government]. 2003. *En omkostningseffektiv klimastrategi* [A cost efficient climate strategy]. February 2003, http://www.fm.dk/Publikationer/2003/En%20omkostningseffektiv%20klimastrategi.aspx?mode=full (accessed 7.6.2010)

Regeringen [The Danish Government]. 2008. *The Danish Commission on Climate Change Policy.* http://www.klimakommissionen.dk/en-US/AbouttheCommission/Sider/Forside.aspx (accessed 7.6.2010)

Regeringen [The Danish Government]. 2009.*The Danish Business strategy on climate change,* November 2009. http://www.denmark.dk/NR/rdonlyres/C14F421F-F33E-4EB4-BF9E-786086019C3F/0/SingleTheDanishBusinessStrtegyonClimateChange.pdf (accessed 7.6 2010)

Späth, P. and Rohracher, H. 2010. 'Energy regions': The transformative power of regional discourses on socio-technical futures', *Research Policy,* 39(4), 449–358.

Sperling, K, Hvelplund,F., Mathiesen, B.V. 2011, Centralisation and decentralisation in Strategical energy planning in Denmark, *Energy policy,* 39, 1338–1351.

Stauning,I., Holm, J., Søndergård B. 2011. Klimaomstilling af byggesektoren – kommunernes rolle. (Climate transition in the building sector – the role of municipalities).CIBS-project report. In Danish.RUC. Available at: http://www.klimabyggeri.dk/links-litteratur.php .

Sønderborg. 2009. *Roadmap 2010–15. ProjectZero for CO₂-neutralt Sønderborg-område* [ProjectZero for a CO_2-neutral Sønderborg-community], http://www.projectzero.dk/lib/file.aspx?fileID=391&target=blank (accessed 1.6.2010)

Voß, J-P. and Kemp, R. 2006. Sustainability and reflexive governance, in *Reflexive Governance for Sustainable Development,* edited by J-P Voß et al. Chelterham: Edward Elgar, p 3–28.

The Development of Non-fossil Energy Systems in the Absence of Strong Climate Change Global Governance

Rikke Lybæk[1], Ole Erik Hansen, Jan Andersen

Introduction

COP 15 held in Copenhagen in 2010 demonstrated the difficulties facing the United Nations in establishing a strong legal framework to counter global warming. This chapter is based on the assumption that these difficulties are primarily due to the limitations of the global policy system and that the discourse on global warming continues to have a very strong momentum in the global community. Thus we expect that many states, companies and civil society actors will participate in the development of new institutions, ways of living and technologies in order to reduce the emission of greenhouse gasses. The concern for the climate is a prime driver for this, but we will argue that this development – in the context of a depressed world economy – is also part of a process of creative destruction (Schumpeter, 1942) offering competitive advantages for states and companies that are first movers. This relates to the possibility of prospective obligations forcing states and other actors to act but also to the fact that fossil fuel is a very limited resource and, in the longer term, the prospect is that oil in particular will be very expensive (IEA, 2007).

Therefore, we expect that, in different institutional settings, there will be a lot of experimentation with the development of technologies and planning and regulation of technologies addressing global warming. It is important, though, that these strategies address the industrial sectors that are most important such as manufacturing, the energy system, the transportation system, the agricultural system and the housing industry. And it is important to learn from experiences with the development of cleaner technologies so as to find out how deliberate policies can stimulate different actors to cooperate to make it possible for new technologies and technological systems to mature.

The authors' background lies in the analyses of different strategies and implementation processes relating to the greening of industries and implementation of non-fossil energy systems in different settings. Of the three authors Ole Erik

---

1 Corresponding author.

Hansen has analysed examples of ecological modernization processes in Western Europe in different sectors to find out how the combination of environmental and industrial policies could trigger the adaptation and diffusion of cleaner technologies in different industrial sectors (for example Hansen et. al., 2002). An important finding was that many 'transitions' were relatively limited in scale and scope because they had to be integrated into the existing technological system and were thus path dependent (Hansen et. al., 2010; Søndergård et. al., 2004). Rikke Lybæk has analysed the possibilities for the implementation of non fossil energy systems in South East Asia (for example Lybæk, 2004; Lybæk, 2009a). And Jan Andersen has analysed energy planning in Denmark and the implementation of non-fossil energy technologies as a part of the Danish energy system.

In this chapter we will combine our knowledge of important (f)actors in relation to innovation and transition processes with our in-depth knowledge of energy planning and the energy system and we will focus on *how deliberative planning processes can stimulate radical transition processes to local non-fossil energy systems*.

The chapter proceeds as follows. In the first part we will discuss important findings from theories on governance of socio-technical regimes. They point to the importance of changes in the 'landscape' – for example the problem of climate change – as important drivers for creating windows of opportunities for novelties. Socio-technical regimes are considered to be dynamically stable and path-dependent. Therefore there is a need for meta-governance and the establishing of 'transition arenas' in order to frame the development of non-standard technologies and learning processes.

In the second part we analyse the Kalundborg Symbiosis. Though it was not originally organized in order to establish a fundamental transition of the local socio-technical regime, but focused on simple gains in water and energy consumption, we find that an arena was established in which the actors engaged in a learning process and developed an even more ambitious project in relation to energy savings and savings in resource consumption. We analyse the learning process in order to find the important (f)actors for the development of the symbiosis; such as conditions connected to stakeholders and networks, conditions composed by the planning frameworks for local resource utilization, conditions related to energy and environmental policies, and finally conditions connected to economic gains for participating industries in order to identify why the symbiosis did not result in more radical shifts in the socio-technical regime.   We argue that the arena has been dominated by incumbent actors representing the dominant local socio-technical regime, and that there have been a lack of meta-governance formulating more ambitious goals for the symbiosis for example the goal of transforming the system in order to establish a non-fossil energy system. Based on this analysis we will make suggestions concerning how to develop such symbioses in the future in order to frame the development of non-fossil energy technologies.

This leads to the third part where we analyse the Navanakorn Industrial Park in Thailand. Our analysis shows that from a socio-technical point of view there are excellent conditions for the development of a local non-fossil based energy system. We identify the factors that can stimulate the establishment of a transition arena and the potential stakeholders that could form the arena. Based on the experiences from the Kalundborg Symbiosis we point to the importance for meta-governance and we identify the obstacles to overcome in order to ensure learning processes.

Governance of Sustainable Transition Processes in a Systemic Perspective

Sustainability is the challenge of our time, and it should be an integral part of contemporary processes of modernization (Hansen et. al., 2010). Inspired by evolutionary and institutional economics, social construction of technology theories and theories of innovation systems, energy and environmental planning has to focus on the management of the transition of socio-technological systems (Geels, 2004; Stærdahl et. al., 2006). Environmental planning has turned into a governance problem: how to stimulate and shape the configuration of the socio-technical systems and how to install a capacity of reflexivity in relation to environmental problems and sustainable development (Stærdahl et. al., 2006).

Environmental concern was integrated into ecological modernization strategies in Europe in the early 1990s (Weaver et. al., 2000; Kemp and Rotmans, 2001; Remmen, 2006). The idea was to combine business opportunities with cleaner technologies. Frequently it was actually possible to establish win-win situations (for example Søndergård et. al., 2003; Holm et. al., 2002; Holm et. al., 2010). Studies show that these environmental innovations was embedded in and shaped by complex patterns of interaction of actors and institutions (Hansen et. al., 2002; Holm and Stauning, 2002). An important finding is that to be successful there is a need for policies stimulating and modulating the configurations of actors and networks to obtain environmental change (Stærdahl et. al., 2006).

The necessity of reflexive governance (Beck, 2006) in relation to sustainable transition to some extent relates to the fact that the concept of 'sustainability' is open to interpretation. Potential transformation paths and effects are highly uncertain because they are a result of complex interactions between social, technological and ecological processes that cannot be fully analyzed and predicted. Goals remain ambivalent because they are endogenous to transformation itself. Conflicts between objectives cannot be resolved scientifically or politically once and for all. And the power to shape transformation is distributed among many autonomous and interdependent actors without anyone having the power to control everyone else (Voss et. al., 2006).

The understanding of technology as socially embedded implies that changes of technologies go together with changes of actors, relations and institutions. This implies a high degree of path-dependency and rigidity making it adequate to talk about socio-technological regimes defined as a dominant actor-network and institutions (Kemp and Loorbach, 2006) with *"dominating practices, norms and shared assumptions, which structure the conduct of private and public actors"*

(Kemp and Rotmans, 2001:7). The regime forms norms and practices that frame processes of innovation and the diffusion of technology. Shifts in regimes include changes in technology, user practices, regulation, industrial networks, knowledge, symbolic meaning (Geels, 2002 and 2004).

The path dependency of socio-technological regimes makes more radical shifts of technologies complicated and often ecological modernization strategies result in innovations of rather limited scope (Søndergård et al, 2003). Thus, Rip and Kemp (1998) and Geels (2002 and 2004) argue that changes must be perceived as multi-level processes with an interplay of the development of niches, regimes and the socio-technical landscape. The landscape includes such elements as material infrastructure, the macro-economy, political culture, social values, belief systems, demography and nature. Niches are local domains where non-standard technologies and new learning processes emerge. Niches can be developed as a bottom up process based on social movements (Smith, 2003) or as a deliberate process of transition management (Kemp and Rotmanns, 2001). Berhhout et. al. (2003) make distinctions between change processes based on internal or external resources and with a high or low coordination.

Socio-technological regimes are complex systems subject to steering attempts by actors inside and outside the regimes. They are the collective – and contingent – outcome of the strategic choices and social interactions of many actors. They defy blueprint steering. Thus the transition of socio-technical regimes for sustainability becomes a question of modulating the governance structures and dynamics of the specific regime, for example by defining long term goals, by economic incentives and by energy and environmental planning. This process of meta-governance of transition management can be described as *"a forward-looking, adaptive, multi-actor governance aimed at long-term transformation processes that offer sustainability benefits"* (Kemp and Loorbach, 2006:103).

A central instrument in turning transition management into operational policy and planning is the formation of a multi actor network (transition arena) in relation to specific sectors and transition goals with the aim of establishing new rationalities and capacities of the innovative system. The major objectives of transition areas are to define problems, establish transition visions and transition goals and to create public support and broaden the coalition (Ibid.). The arena should form the basis of a reflexive governance process that in a cyclic way moves from stages of organizing multi-actor networks, developing sustainability visions and transition agendas, mobilizing actors, executing projects and evaluating, monitoring and learning. It should condition the process of suggesting alternative technologies and facilitating and modulating the generation of a variety and technology options and selection processes (Stærdahl et. al., 2006).

The establishment of such meta-governance raises difficult questions about who is going to identify and install actors in transition arenas (Kemp and Loorbach, 2006). The transition management approach has a tendency to be too optimistic regarding the capacity of the state to manage both the formation of the transition arenas and to establish a framework for transition processes.

We are going to examine this question in more detail in our analysis of the proposed transition of an energy system in Thailand, but at a general level we will argue for the necessity of forming a meta-governance network involving actors from companies, the local community, local authorities and state agencies.

Furthermore, transition arenas are always situated in specific contexts and they are very often related to established technology systems. There is a high risk that they are captured by incumbent players and subject to their strategic interests, that they reproduce established paths, stick to less radical alternatives or pick up suboptimal technologies and development paths (Hisschemöller, 2006). This all points to the importance of a form of meta-governance that can define new goals for the development of transition processes and for the establishing of competing transition arenas.

Transition in practice: Kalundborg Symbiosis, Denmark

Within the framework of interfirm collaboration the notion of Industrial Symbiosis (IS) has emerged. Chertow (2004) defines IS as follows:

> "Industrial symbiosis engages traditionally separate industries in a collective approach to competitive advantage involving physical exchanges of materials, energy, water and/or by-products. The key to industrial symbiosis is collaboration and the synergistic possibilities offered by geographic proximity" (Chertow, 2004:2).

In the west Zealand municipality of Kalundborg, IS collaboration takes place between industries in the area. Kalundborg is located approximately 100 km. from Copenhagen and has 20,000 residents. The Kalundborg Symbiosis is a paradigmatic example of symbiosis and evolves collective resource optimization based on by-product exchanges and the sharing of energy between the different industries placed in this specific geographic area (for a detailed introduction see for example Ehrenfeld and Gertler, 1997; Côte and Cohen-Rosenthal, 1998; Esty and Porter, 1998; Ehrenfeld and Chertow, 2002; Jacobsen and Anderberg, 2004; Jacobsen, 2006).

The motivation for materials and energy exchanges arose in the early 1970s from a mutual effort to reduce costs through income generating use of "waste" products. Gradually, the industries involved realized that exchanges of resources could enable mutual economic benefits, and at the same time reduce the environmental impact of large industrial manufacturing. Today, Kalundborg's Industrial Symbiosis comprises the following eight core industries (six processing): DONG Energy's Asnæs CHP plant, Statoil-Hydro Refinery, Gyproc, Novo Nordisk, Novozymes, recycling company RGS 90, waste company Kara Noveren and the Kalundborg Municipality. Each industry is connected through different types of network relations; flows of steam, gas, water, gypsum, fly ash and sludge etc. (The Kalundborg Centre for Industrial Symbiosis, 2010).

Presently Asnæs CHP plant produces heat for the city of Kalundborg through the supply of district heating, and also supplies process steam to Statoil Refinery, Novo Nordisk and Novozymes. The combination of heat and power production (CHP) results in a 30 % improvement in the fuel utilization compared to a separate production of heat and power. Approximately 4,500 households in Kalundborg receive district heating from Asnæs CHP plant. This heat supply replaces approximately 3,500 individual small oil-fired boilers (Ibid.)

Statoil Refinery receives process steam and water from Asnæs CHP plant. The steam covers about 15 % of the refinery's total consumption of steam. The refinery uses the steam for heating oil tanks, pipelines and so forth. Novozymes and Novo Nordisk use steam from Asnæs CHP plant for the heating and sterilization of the processing plants. Moreover, salty cooling water from Asnæs CHP plant is used by a local fish farm producing 200 tonnes of salmon yearly. The fish have better growth conditions in the heated water (Ibid.).

Gyproc receives excess gas as energy input from Statoil-Hydro Refinery and industrial plaster as materials input from Asnæs CHP plant. Calcium and recycled treated waste water are added to the sulphur extracted from the flue gas at Asnæs to form industrial plaster, thus gypsum. Insulin production at Novo Nordisk releases materials which are distributed to surrounding farms as pig fodder. A by-product of the yeast fermentation process is converted into yeast slurry, which replaces approximately 70 % of the soy proteins in traditional feed mixes. Novo Nordisk adds sugar, water and lactic acid bacteria to the yeast in order to make it more attractive to the pigs (Ibid.).

The Kalundborg Symbiosis has developed spontaneously over a number of decades and comprises 20 different projects and it is still evolving (Ibid.). The materials and energy exchanges comprising the symbiosis are outlined in Figure 2.3.1:

(F)actors of importance for the establishment of and continuous evolution of the Kalundborg Symbiosis

The Kalundborg Symbiosis is a well analysed case and based on these analyses we will discuss the *conditions* for the formation of and continuous evolution of the Kalundborg Symbiosis; thus we ask which **networks and stakeholders** (interfirm relations) have provided the necessary platform for a development of the symbiosis? And which **energy and environmental policies** (CHP & standard on air emissions) and **planning frameworks for resources utilization** (water & steam) can be identified? Finally, how has the industries **economy gains** been affected by the symbiotic activities?

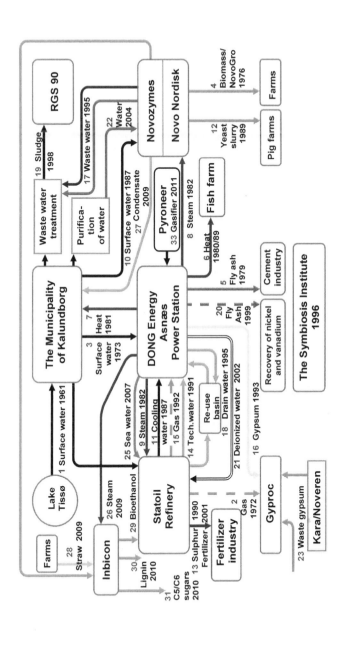

Figure 2.3.1 The Kalundborg Symbiosis

Source: The Kalundborg Symbiosis Institute (2010)

Network/Stakeholders (interfirm relations)

Business culture: Mutual trust among the stakeholders participating in the Kalundborg Symbiosis has been essential for the evolution of the symbiosis. The industries have been open to exposing their internal weaknesses in relation to materials and energy consumptions, and welcoming outside suggestions and advise from various stakeholders.

Small community: The relatively small community of Kalundborg has also facilitated the interfirm collaboration between industries, as managers and CBOs from the participating industries have been engaged in the same social circles in the community. This has helped to sustain the mutual trust emphasised above, and the opportunity to constantly exchange ideas and visions for the development of the co-operation (Pers. com. Christensen, 2009).

Bottom-up: The development of the symbiosis has thus evolved as a genuine bottom-up process initiated by the stakeholders themselves and not as a result of governmental requirements and regulations (at least not in the initial phase).

Thus, the *conditions in relation to network/stakeholders* have been ideal in the case of Kalundborg, providing a platform consisting of mutual trust in a relatively small community enabling a genuine bottom-up process to evolve.

Planning framework (water & steam)

Water: The symbiosis was originally initiated due to water scarcity and costliness of groundwater resources, and then continued to evolve, not only including water, but also energy and materials. The optimization of and better access to natural resources, in this case water, were thus the starting point for the interfirm collaboration (Jacobsen, 2006). This was gradually achieved by replacing industrial consumption of groundwater with surface water and establishing a reservoir for water recycling, by optimizations of internal water uses within the industries, and by upgrading surface water to drinking water quality for the citizens of Kalundborg (Ibid.). Thus, the *planning framework conditions in relation to water* have primarily been lack of access to affordable groundwater resources, as well as very large options for applying water savings.

Steam: The location of Asnæs CHP plant in a scarcely populated part of Zealand has also created unique symbiotic possibilities, as huge amount of surplus heat/steam became available when the plant was converted from a power plant to a CHP plant during the 1980s. Thus, the small city of Kalundborg was not able to consume this amount of energy supplied by means of district heating, and the pipeline distances to other larger cities were too far.

Therefore, various initiatives were established to make use of this surplus heat/steam. As mentioned, earlier steam from the CHP plant is, for example, used at Statoil, mainly for cleaning oil tanks and pipelines, and for heating up boilers. At Novozymes and Novo Nordisk, the steam is used for heating and sterilization purposes. A final step in the utilization of heat/steam is the utilization of surplus heat in the nearby fish farm. After the production of power, steam and district heating, the boiler water from the Asnæs CHP plant is condensed with salty cooling water. The cooling water then reaches a temperature of 7 to 8 degrees C. above normal (Jacobsen, 2006), which is very suitable when raising the approximately 200 tons of salmon fish produced on a yearly basis.

Thus, the steam related symbiosis was – as opposed to the water symbiosis – not initiated with the specific purpose of optimizing access to resources, but simply motivated by options for economic gains. It should however be mentioned that despite the initiatives for reusing the waste heat/steam from Asnæs CHP plant for economic gains, the final total amount of excess energy (hot salty cooling water) being discharged in 2002, added up to 1,560 TJ (or 1.56 PJ). Of this the fish farm utilizes only approximately 39 TJ (or 2.5 % of the total amount of waste heat/steam) (Jacobsen, 2006).

Thus, the *planning framework conditions in relation to steam* have primarily been the huge amount of surplus heat/steam enabling a search for re-use options, in combination with a nearby 'market' for distribution of surplus heat/steam composed by the surrounding industries' demand for heat at different temperatures.

Energy and environmental policies (CHP & air emission standards)

Energy and environmental related concerns also played a role in the evolution of the Kalundborg Symbiosis. Apart from governmental energy efficiency requirements related to the conversion of Danish power plants to CHP plants introduced during the 1980s – resulting in large quantities of surplus heat/steam being available in Kalundborg – another important environmental regulation also influenced the further development of the Kalundborg Symbiosis as will now be shown.

The implementation of the wet flue gas desulfurization plants at Asnæs CHP plant in 1992 and again in 1995/1996 – reducing the SO_2 emissions to the atmosphere by removal from the flue gas – initiated an even higher demand for water at the energy facility, as well as the development of a new symbiosis. The first desulfurization plant established in 1992 required approximately 450,000 m3/year of water to operate, and the second plant, another 400,000 m3/year (Jacobsen, 2006). This massive requirement for additional water supply for operating the desulfurization plants hence developed the water-related symbiosis even further during this period. Thus, environmental regulation reinforced the interfirm collaboration.

The implementation of the desulfurization plants also led to the establishment of Gyproc, a company manufacturing planter boards on the basis of gypsum. The gypsum collected from the flue gas of Asnæs CHP plant reduces the import of natural gypsum. It is more uniform and pure compared to natural gypsum, which makes it appropriate for construction materials (Gullev, 2005). Thus, the *conditions in relation to energy and environmental policies*, exemplified here, have been higher energy efficient standards for energy utilities (CHP requirement), and well as SO_2 emission reductions providing a better air quality standard.

Economy gains

The Kalundborg Symbiosis has, over the years, resulted in economic gains (or saved expenses) for the industries involved. But looking at the total amount of materials and energy throughput within the industries participating in the symbiosis, these savings are relatively limited. Huge amounts of resources and money have thus been spent on developing an interfirm cooperation, which basically leads to minor materials and energy savings (Jacobsen, 2006). As such, the Kalundborg case does not provide any transformation of the energy supply system; it evolves around and seeks to optimize on waste (materials and energy) from the biggest fossil fuel CHP plant in Denmark. Thus, the *conditions in relation to economy gains* are quite limited when compared to the actual amount of materials and energy flows passing through the industries participating in the symbiosis, and the transformation of the energy supply system is therefore relatively limited.

Transition in Planning: Navanakorn Industrial Park, Thailand

Navanakorn is an industrial park located in Pathum Thani Province, Takhlong Municipality, 45 km. north of Bangkok, Thailand. There are approximately 250 industries located in Navanakorn distributed within four industrial zones in the park (Navanakorn, 2008), and the number is growing as new industries constantly establish themselves in the park. There are many types of industries located there: food, medical and furniture industries, electronically and assembly industries, etc. Few of them are large scale businesses, i.e. having more than 2,000 employees, but can in general be characterized as small and medium sized enterprises (SMEs) with less than 500 employees (Lybæk, 2004; Navanakorn, 2008).

Close to 40 % of the industrial output in Thailand measured by GDP is produced by around five million such SMEs primarily located within approximately 80 Industrial Parks (Pongvutitham, 2010). These industries therefore account for a significant amount of the total energy consumption of the industrial sector. Within this sector 'manufacturing' is one of six sub-sectors (agriculture, mining, manufacturing, construction, residential, commercial and transportation sector), which account for 37 % of the total energy consumption in Thailand, or 24,195 ktoe (thousand tonnes of oil equivalent). The energy types consumed in this sector are primarily coals, renewables, electricity, petroleum products, and natural gas (DEDE, 2008).

SMEs in the wood, pharmaceutical and food manufacturing lines of businesses are selected as case industries inside Navanakorn, as a relatively high percentage of their energy demand is hot water rather than steam (thus below 100 degrees C.). On average, 65 % of the heat demand in wood, pharmaceutical and food manufacturing SMEs requires hot water and not steam (Lybæk, 2004). Thus, conversion to low temperature district heating is a genuine possibility.

The present situation (existing energy system) for the six SMEs case industries, in the food, wood and chemical line of business, presented here, is that they all rely on supply of fossil fuels in the form of power from centralized Thai CHP plants transmitted on the national grid. They also rely on outside supply of fuel oil (bunker oil) for heat-only production in individual boilers (except from Rockwood who produce process heat by means of power).

The efficiency of the energy supply system is relatively low, as it is based on a separate production of power and heat (heat-only boilers and power supplied from the national grid). The overall efficiency of power plants implemented in Thailand is also quite low (Greasen and Footner, 2006). There are no options for industries to be connected to a joint energy supply system (as for instance a district heating network) within the park, thus all businesses within Navanakorn cover their supply of heat individually.

Valuable resources are moreover transported out of Navanakorn Industrial Park, as the collected industrial biomass waste from within the park, primarily goes to landfills or to other outside sectors for inefficient re-use, which leads to intensified transportation, hence emissions of CO_2 and methane (CH_4). Today, other provinces in Thailand receives the landfill waste from Pathum Thana, as only limited spatial areas can be found for waste disposals in this province (Lybæk, 2004).

The future situation (transformed energy system, by means of a biomass CHP with supply of district heating) suggested in Figure 2.3.2 seeks to exploit industrial biomass resources from within Navanakorn, and minimize the flow of unsustainable materials (fossil fuel based energy) *into* the park. It also seeks to prevent valuable resources being discharged or re-used inefficiently outside the industrial park. The energy system suggested thus rely on a local self-supply scheme, in which local biomass waste are used as fuel in a "transformative" CHP technology, which distribute power and heat to *several* (here six) SMEs connected to a small scale district heating network. Local agricultural biomass waste, or other sources of clean organic waste from the community, can also be supplied in order to increase the amount of biomass fuel.

The district heating network is indicated in Figure 2.3.2 as a bold circle from which the industries extract heat by means of heat exchangers through individual service lines, substituting fuel oil usage. As consumption of steam now only happens at Imperial and to a minor extends at B.B. Snacks, due to technical adjustments, it is now possible to cover the process heat demands in the remaining industries solely by district heating (hot water not steam). This means that energy savings are obtained both qualitatively (from 'steam' to 'hot water') and quantitatively (by the reduced amount of energy which it now is necessarily to produce in order to cover the energy demands). The sustainable energy supply system proposed below is not implemented in Thailand, but suggested on the background of thorough empirical analysis and studies, etc. (see Lybæk, 2004; Lybæk, 2008; Lybæk, 2009a; Lybæk and Jacobsen, 2009b; Lybæk and Andersen, 2010).

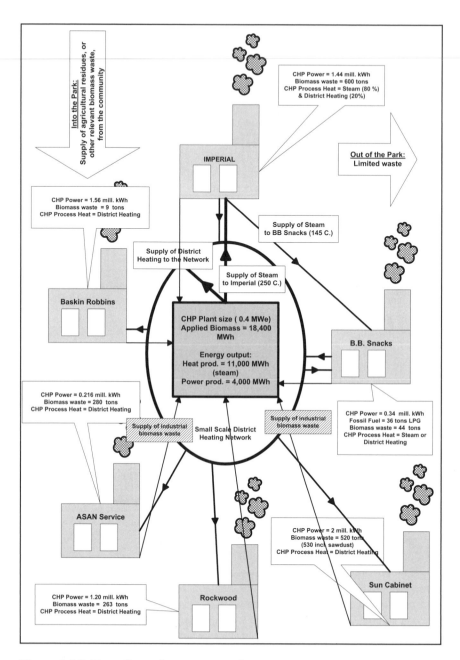

Figure 2.3.2 Transformed energy supply system proposed in Navanakorn Industrial Park

Source: Own figure and background data collection

(F)actors of importance for the implementation of the proposed project in Navanakorn Industrial Park

We will now discuss the *conditions* for the implementation of the proposed project in Navanakorn Industrial Park and ask: Can local **networks and stakeholders** (inter-firm relations), provide the necessary platform for the development of a symbiosis? And which **energy and environmental policies** (CHP and standard on air emissions) and **planning frameworks for resources utilization** (efficient use of resources) can be identified? Finally, how has the industries' **economy gains** been affected by the symbiotic activities? The experiences from Kalundborg will be used to identify (f)actors of importance – thus analogues – for an actual implementation of the proposed project (biomass CHP with supply of district heating to industries).

Network/Stakeholders (interfirm relations)

As seen in the Kalundborg case, mutual trust between and openness among the stakeholders have been two of the reasons why the symbiotic relationship initially emerged. The fact that the community in Kalundborg is relatively small and decision-making people thus moves in the same social circles, have also had an impact. Turning to the Navanakorn Industrial Park this situation is not at all the case, on the contrary.

There is no communication between the industries located in Navanakorn Industrial Park, and the industries keeping company data to themselves. There is no business culture providing a platform for exchanges of knowledge regarding materials and energy uses, or other types of sharing industry 'sensitive' data (Lybæk, 2004). Certain companies in the same line of business, e.g. wood industries will however meet, but in a fora established under the Federation of Thai Industries (FTI). Thus, among the many different types of industries located in the park, there is no communication at all (Ibid.).

Even though the community of Navanakorn Industrial Park is relatively small, industry managers or CBOs live outside the park, and thus only enter the park when working. Thus, no sharing of social circles exists, these could provide a platform for mutual trust and thereby initiate ideas and strategies for interfirm cooperation (Ibid.).

The bottom-up process which evolved in the Kalundborg case is not likely to happen within Navanakorn Industrial Park, due to lack of mutual trust. In our point of view, an *initial* top-down approach must however assist and facilitate the implementation of the proposed energy system (biomass based CHP with supply of district heating). This process is however not intended to be the classical understanding of a top-down approach, in which initiatives taken by the central administration are adopted and implemented down along the hierarchical structure of the society.

A top-down approach is here interpreted as a mix of stakeholders at different levels taking the initial lead and initiative to promote the project ideas, through for instance setting up a show case/demonstration plant, or/and by disseminating information to the local industrial community. After such initial work and promotion of the proposed project, it is our belief that projects like this will emerge within industrial parks in Thailand without any such top-down initiatives, but primarily facilitated by relevant local stakeholders. Here an important stakeholder is already mentioned, namely the FTI who is regarded as a trustworthy stakeholder by the industries, having sub-chapters throughout Thailand.

Table 2.3.1   Network/stakeholder conditions

Kalundborg Symbiosis, Denmark	Navanakorn Industrial Park, Thailand
Mutual trust between industries	No culture of mutual trust between industries
CBOs a part of the same social circles in a small community	No common social circles (except those established by FTI)
A bottom-up process evolved	An initial top-down process required

Source: Own table

Planning framework (efficient use of natural resources)

As identified earlier, lack of access to groundwater resources has been an important factor initiating the Kalundborg Symbiosis. The costly utilization of this limited resource has thus initiated various means of re-use strategies for water in the symbiosis area. An analogy to the utilization of a limited resource can be identified in Navanakorn Industrial Parks, as all case industries make use of fossil fuel (bunker oil) in individual boilers for process heat generation. The power consumption is supplied from the national grid and is also highly fossilized, (primarily generated from oil, coal and natural gas power plants). The cost of this energy supply (purchase of power and bunker oil) has increased dramatically in recent year, and is becoming a serious issue for the industries in Navanakorn (Lybæk, 2004; Lybæk, 2009a).

Different options for applying water savings has been identified in the Kalundborg case. When looking at the options for resource optimizations in Navanakorn Industrial Park, similar initiatives could be applied. This can be more efficient use of the natural resources for energy production by means of for instance CHP, substituting the heat-only production, and a more efficient consumption of the energy within the industries. The latter could be achieved by converting a steam demand to a hot water demand requiring less initial energy to produce.

Thus, the *planning framework conditions in relation to natural resources* are a high utilization rate of costly limited resources as fossil fuels, but optimizations in the resource utilization is possible.

In the Kalundborg case a huge amount of surplus heat/steam enabled the industries to search for alternative re-use options. An analogue to this can also be identified in Navanakorn Industrial Park, where large amounts of excess resources are available. Large quantities of organic industrial and household waste are discharged on a daily basis; transported out of the park to landfills or to inefficient re-use elsewhere. Within the community surrounding the industrial park agricultural residues are also left to decompose on or openly burned on the fields. Alternative re-use of these industrial and agricultural biomass resources are thus a real opportunity.

In Kalundborg the nearby 'market' for distribution of surplus heat/steam, composed by the surrounding industries', helped limit the discharge of surplus energy from Asnæs CHP plant.

In Navanakorn Industrial Park a huge market for distribution of heat is composed by the industries situated in the park. Within these industries there is a high demand for process heat, which presently is covered by small individual fossil fuel boilers for heat-only production.

Thus, the *planning framework conditions in relation to natural resources* are that large amounts of biomass waste are being discharged. At the same time there is a high demand for process heat within industries located in the park, enabling new re-use strategies for presently wasted resources.

Table 2.3.2   Planning framework conditions

Kalundborg Symbiosis, Denmark	Navanakorn Industrial Park, Thailand
Lack of access to groundwater resources (natural resource)	High utilization rate of costly limited resources as fossil fuels (natural resource)
Very large options for applying water savings	Optimizations in resource utilization possible
Huge amount of surplus heat/steam enabling a search for re-use options	Large amounts of biomass waste (industrial & agricultural) now being discharged
Existence of a nearby 'market' for distribution of surplus heat/steam composed by the surrounding industries' demand for heat at different temperatures	High demand for process heat within industries, presently covered by small individual fossil fuel boilers

Source: Own table

Energy and environmental policies (CHP & air emission standard)

In Kalundborg, energy efficiency requirements for energy utilities (CHP), as well as emission standards on air quality (desulfurization plants), pushed forward the continuous evolution of the symbiosis and thus the interfirm collaboration.

Implementing a more energy efficient production of energy at the single industrial level in Navanakorn Industrial Park would only be feasible within large industries requiring high quantities of steam, which then additional could run a steam turbine for power generation, thus CHP production (e.g. at Imperial). Within the remaining industries with a limited demand for steam this is not feasible, but energy efficiency could be applying by using more modern and efficient boilers for heat-only production.

Energy efficiency at the industrial level faces some governmental support in Thailand. Energy efficiency (EE) and Demand Side Management (DSM) within Thai industries are expected to provide 3,190 ktoe of energy savings within 2011, with 7,820 ktoe being the overall target. One third of this industry target is thus expected to be achieved through the implementation of industrial CHP (Opatvachirakul, 2009). In general, however, Thailand has primarily chosen to develop its energy sector by means of larger centralized fossil fuel power plants, even though smaller decentralized CHP plants are an opportunity for Thailand (Lybæk, 2009).

In 2007, Thailand received applications from a large number of Small Power Producers (SPPs) amounting to 2,400 MW of CHP. This amount exceeded the allowed 500 MW set by the Thai government. Later on 740 MW of CHP were however accepted, together with four centralized fossil fuel power plants from Independent Power Producers (IPP) amounting to 4,400 MW. Thus, emphasis on energy efficiency in the industrial sector in Thailand exists, but the Thai governments' focus on adapting to CHP technologies are unfortunately relatively limited. This is however not the case when looking at the actual options for implementation, and the interests from SPPs (Lybæk, 2009).

In the Kalundborg case, environmental requirements on air cleaning equipment lead to the evolution of further symbiotic development among industries. In Thailand, enhanced air quality standards would not necessary lead to interfirm collaboration like in Kalundborg, but more likely lead to fuel switch or adaptation to more efficient boilers, etc. at the single industrial level. The bunker oil which industries normally utilize to generate process heat in Thailand has a relatively high content of sulphur dioxide (SO_2) (Lybæk, 2004). The Thai environmental regulation however has a high emission standard on air pollution from the industrial sector (see PCD, 2010), but the actual control and enforcement of this regulation is weak.

Thus, the *conditions in relation to energy and environmental policies* are relatively weak in Thailand, as the policies on energy efficiency, hereunder CHP, and control and enforcement of the industrial air pollution do not initiate environmental friendly changes.

Table 2.3.3   Energy and environmental policy conditions

Kalundborg Symbiosis, Denmark	Navanakorn Industrial Park, Thailand
Energy efficiency requirements for energy utilities (CHP)	Energy efficiency and CHP technologies has relatively limited governmental focus
Higher air quality standards on SO_2 emissions	Air emission standards are in general high, but the control and enforcement are weak

Source: Own table

Economic gains

Compared to the Kalundborg case, the proposed project in Navanakorn Industrial Park will have a substantial impact on the economy of the participating industries. This is first caused by substitution of costly fuel oil for heat-only boilers with own or community based biomass resources as fuel, second by substitution of costly power from the national grid with own CHP power, and third by saved expenses on waste handling costs etc. (Lybæk, 2004). Looking at the total amount of materials and energy throughputs within the industries proposed to join the biomass based CHP with district heating the savings achieved are thus quite extensive. The proposed energy system is therefore highly transformative in its nature, as it both dematerializes and decarbonizes the local energy supply system.

The estimated economic gains by applying the proposed energy system are outlined in the following, focusing on the 1) cost/benefits by applying the system, and on the 2) potential jobs connected to it.

The yearly economic contribution from the system is calculated to 2.7 million Danish Kroner, (DKK) on the basis of a cost/savings analysis. Thus, compared to the existing system with purchase of fuel oil and power from the grid, operation costs of individual boilers and expenses for waste management handling, etc., the industries save or gain 2.7 million DKK per year. This is achieved after setting up the small scale biomass CHP plant with district heating pipes, and using the industries' own biomass waste as fuel instead of discharging it (for more thorough details, also on the economic methodology applied in the above calculations, see Lybæk, 2004).

The jobs connected to the project can be divided into 'construction' and 'M&O' ones. The construction jobs are short term jobs connected to the building of the plant. The specific type and size of the plant will thus lead to an estimate of 10 construction jobs (more details are outlined in Lybæk, 2009a). Jobs related to maintenance and operation (M&O) are long term jobs connected to for instance fuel supply and technical operation, etc. Again, the type and size of the plant are used as an indicator for the potential jobs, which are found to be approximately seven. Thus, on top of the economic contribution for industries involved in this interfirm cooperation, approximately seven local jobs can be created, on the background of a relatively small project.

Table 2.3.4   Conditions for economy gains

Kalundborg Symbiosis, Denmark	Navanakorn Industrial Park, Thailand
Economic savings limited when looking at the amount of materials and energy passing through the industries	Options for economic savings are high as affecting a large share of the materials and energy passing through the industries
Limited impact on transformation of the energy supply system	Large impact on transformation of the energy supply system

Source: Own table

Thus, the *conditions in relation to economy gains* are very favourable as the options for economic savings are high, as the proposed energy supply system affects a very large share of the total materials and energy flows passing through the industries. This again means that a large transformative impact regarding the energy supply system can be achieved.

Environmental benefits

Apart from the economic gains outlined above there are also environmental benefits connected to the proposed energy system in Navanakorn. The CO_2 emissions will be reduced from 10,413 tons/year to neutral, NOx lowered from 12.4 to 4.4 tons/year (64.5 % reduction), and SO_2 from 21 to 2 tons/year (90.6 % reduction) (Lybæk, 2004). The two latter will benefit the local community, as the air quality improves significantly.

Discussion of Actions Supporting Symbiotic Activities in Thailand (Stakeholders, Activities, Regulations etc.)

In the following section some examples of actions to support symbiotic activities in Thailand will be outlined. Many more actions than those can be proposed, which are explored more thoroughly in Lybæk, 2004 and Lybæk, 2009a.

Actions related to Network/Stakeholders

In the initial face of promoting the proposed project in Navanakorn (and in Thailand in general), a team of leading stakeholders must be created to commence this development and to disseminate the concept to industries located within industrial parks in Thailand (an *initial* top-down approach). As already mentioned one member should be the Federation of Thai Industries (FTI), as it is a powerful and trustworthy stakeholder with the capacity to build bridges between industries. First, by promoting the idea in Navanakorn across different lines of businesses, and second, by the already established sub-chapters to FTI throughout Thailand. Other important stakeholders in the initial phase would be a power full NGO such as the Thai Environmental Institute (TEI) or/and academia representatives, e.g. Asian Institute of Technology (AIT). This team of leading stakeholders should disseminate the proposed project to owners and managers of the industrial parks to the province and the municipality in which the parks are located, etc. (Lybæk, 2004).

The tasks of the leading stakeholders are, thus, to create an interest among industries to engage in symbiotic activities by promoting the proposed project. This can be done by campaigning and awareness rising activities. This could be in the form of written material or web sites presented to industries, as well as factory visits by the team of leading stakeholders. The information can contain economic and environmental benefits for industries when joining such project. Later, workshops and seminars can gather potentially interested industries for further information and discussions. The most important aspect when looking for participants to join the symbiotic activities is campaigning and awareness rising activities, and through these to identify industries who will participate on a voluntary basis (Lybæk, 2004).

Actions related to Economy gains

To further disseminate the economic gains for industries engaging in symbiotic activities, here by means of the proposed project, it would be very fruitful if a show case/demonstration plant could be established (for instance in Navanakorn Industrial Park). This would not only illuminate the economic benefits (saved expenses on fuel oil, power purchase, waste handling costs, etc.), but also prove that inter-firm cooperation provides a number of additional benefits for the industries evolved and for the local community, as a whole. Apart from the saved expenses etc. already outlined, new business opportunities could also emerge. These might include options for sale of power and heat to other industries within the park, sale of ash as farmland fertiliser and the opportunity for local farmers to supply agricultural residues as fuel to the energy facility. Apart from that, the local authorities (the provinces and municipalities) will be able to supply certain types of waste to the energy facility, thus limiting the need for landfill development (Lybæk, 2009).

In the case of Navanakorn, such a showcase could be co-financed by the owners of the park, by some of the industries participating in the proposed project, and by grants from the Thai government. Tax exemptions, or other types of economic benefits for industries participating in such projects, could also be provided by the local authorities, thus supporting the development of interfirm cooperation which includes waste minimization initiatives (Lybæk, 2004).

Actions related to the planning framework

To be able to design the most appropriate energy supply system in the given context, it is important that thorough investigations of the materials and energy throughputs within industries located in the parks are investigated. Thus, the amount and type of industrial biomass waste must be identified, as well as other resources from within or outside the industrial parks (sludge from waste water treatment plants, waste from household and commerce, agricultural residues, etc.). Also, the amount and type of energy usage, and the type of processing equipment and options for improving the energy efficiency, are important knowledge. The latter especially in relation to the conversion from steam to hot water demands (supplied by district heating) within industries.

Thus, based on the knowledge of the specific energy *demands* within industries, and the options for applying energy efficiency, as well as the biomass resources available, an actual design of the energy supply system can be outlined.

Relevant stakeholders would here be Department of Industrial Works (DIW), due to their knowledge on waste and waste handling activities within industries, as well as the Ministry of Energy (MOI), more precisely DEDE (Department of Alternative Energy Development), in relation to energy usage and technical issues in general. They can be assisted by King Mongkuts Institute of Technology (KMIT), or the before mentioned AIT and TEI. As FTI has the capacity to conduct all the analysis, they would be the lead stakeholder in these investigations (Lybæk, 2004).

Actions related to energy and environmental policies

Energy and environmental policies in Thailand supporting the development of symbiotic inter-firm cooperation should not be based on a top-down approach, but on more facilitating principles. If any laws or regulations should be applied, it should only be with the purpose of breaking down existent barriers for such cooperation. As far as the latter issue the Thai regulation presently require a long and difficult application period before allowing industries to move and thus exchange their wastes. A quick approval process would here help industries to engage in inter-firm cooperation (Lybæk, 2009).

Regarding issues related to facilitating principles, the Thai government already supports the use of biomass waste for energy purposes through a feed-in electricity tariff, which has been quite successful. A feed-in tariff could however be developed focusing on the actual *use of generated heat/steam*. The co-production of heat and power tend to be quit inefficient in Thailand, as the market for heat is limited. It is often the industry itself (the rice-mill or palm oil mill, etc.) that constitutes the heat 'market'. With the favourable feed-in tariffs on power, the industries has focused on sale of power to the grid and discharged any surplus heat which they could not export or use themselves.

Within the industrial parks on the contrarily a large market for export of surplus heat/steam exists, constituted by the industries, which could be exploited by support from a new type of feed-in tariff. Such a feed-in tariff would thus help efficient renewable energy technologies (CHP) to be implemented where there is an actual demand for heat, and where large quantities of biomass residues are present. This will also result in a much more efficient utilization of the biomass waste, compared to presently applied practices. An important stakeholder in setting up this support is MOI.

Concluding Remarks on Meta-governance and Stakeholder Networks

We have tried to find out how deliberative planning processes can stimulate radical transition processes to local non-fossil energy systems. Inspired by theories on reflexive governance in relation to sustainable transition, we have analyzed transition processes in the Kalundborg Symbiosis in order to propose how transitions processes to non-fossil energy systems can emerge in an industrial district in Thailand; Navanakorn Industrial Park.

It is important to be aware of some important differences between the two examples. Kalundborg is an example of a systemic transition of a local waste and energy system. It was *not* sustainability driven, and there was no meta-governance in order to establish a transition arena. Meta-governance – especially national priorities in relation to the energy system and waste management – was important, but the first driver was limited water resources. Later on economic gains from the re-use of surplus heat/steam was an important driver.

In Thailand we argue for the importance of a deliberative meta-governance, in order to set up incentives and to establish a transition arena. Sustainability is an important driver in the sense that the transition process relates to national and global priorities, focusing on the reduction of waste and emissions of CO_2.

In Kalundborg, the transition arena was dominated by incumbent players in the local socio- technological system. We argue that the transition process was not in all cases subject to but in accordance with their strategic interests. There was a focus on a more efficient use of limited resources, but actually it is an important consequence that the coal based CHP plant is very difficult to close down as it is the main cornerstone of the symbiosis. Thus, Kalundborg is an example of the path dependencies established in transition processes. This point to the need for the establishing of competing transition arenas, where non-standard technologies such as non-fossil based energy systems is developed.

In Thailand we propose the establishing of a transition arena with an actor network, where the incumbent players in the local socio-technological system have an important role. Based on the experiences from Kalundborg this may in the long run cause problems, because these actors have the possibility to capture the definition of sustainability. On the other hand, it is an advantage that they are not incumbent in Thailand's fossil based energy system. Reduction of CO_2 emissions is presently an important global priority, but it is vital to learn from the perspective of reflexive governance and integrate actors that can stimulate reflections on the interpretation of sustainability. In this way the important social or environmental dimensions are not excluded. Representatives from the local community and actors with knowledge on sustainable production could be crucial participants in the local network. In the beginning it is not important to organize competing transition areas within the industrial district. But it is critical that the experiences from Navanakorn are communicated to other industrial districts throughout Thailand in order to establish similar transition arenas'.

An important lesson from Kalundborg concerns the significance of the specific project that gives short term advantages for the involved actors. Therefore, in Navanakorn we suggest an organization of the transition process around a show case/demonstration plant offering immediate economic and environmental benefits. This hard core planning conflicts to a certain degree with 'reflexive governance', but we will argue that the important thing is to establish a transition arena and to develop trust among the actors. These benefits will give an incitement for the local actors to involve themselves in additional transition processes.

In Thailand we argue for a deliberate process of transition management as it has not been possible to identify the social movements which could initiate the transition process as a bottom-up process (Lybæk, 2004). It is important, however, that the transition arena is organized in a way that facilitates bottom-up processes in the transition process. The establishing of a show case/ demonstration plant is thus an important starting point, but the transition arena has to be open for different perspectives for moving towards sustainable development in the local energy and production system.

Both in Kalundborg and in Navanakorn local resources play an important role. At the same time, though, it is important in Thailand to facilitate an adoption process where technologies are integrated in the local energy and production structure. It is vital that this is not just treated as transfer of technologies. It has to be adopted and the local knowledge and ownership is thus imperative in order to make the technology an integrated part of the local systems.

The important lesson from Kalundborg is that pragmatic projects resulted in new constellations of actors with shared assumptions, and new relations, practices and institutions. We have identified some obstacles for such a development in the Thai context, but we will argue for the importance of external competencies and a high degree of coordination; especially in the beginning of the process. Economic benefits are very important to start the process, but within a few years Navanakorn Industrial Park might be a show case for sustainable transition strategies in relation to local energy and production systems, as an example of sharing and exchanging resources. This could attract new actors thereby making the development more sustainable.

References

Beck, U. 2006, Reflexive governance: Politics in the global risk society, in Voss, J. P., Bauknecht, D., Kemp, R. (eds.) Reflexive Governance for Sustainable Development, Edward Elgar, Chelterham, 31–56.
Berkhout, F., Smith, A., Stirling, A. 2004, Socio-technological regimes and transition contexts, in Elzen, B., Geels, F. W., Green, K. (eds.), System

Innovation and the transition to sustainability- Theory, Evidence and Policy. Edgar Elgar, Chelterham UK, 48–75.

Chertow, M. 2004, Industrial Symbiosis, in Encyclopedia of energy, edited by C.J. Cleveland. Oxford. Elsevier.

Coté, R. and Rosenthal E. C. 1998, Designing eco-industrial parks: A synthesis of some experiences. Journal of Cleaner Production, 6 (3–4) 181–188.

Department of Alternative Energy Development and Efficiency, DEDE, 2008, Thailand Energy Situation 2008, Thai Ministry of Energy (MoI).

Esty, D. and Porter, M. 1998, Industrial Ecology and competitiveness: Strategic implications for the firms, Journal of Industrial Ecology, 2 (1) 35–43.

Ehrenfeld, J. and Gertler, N. 1997, Industrial Ecology in practice: The evolution of interdependence at Kalundborg, Journal of industrial Ecology, 1 (1) 67–79.

Ehrenfeld, J. and Chertow, M. 2002, Industrial symbiosis: The legacy of Kalundborg. In Handbook of industrial ecology, edited by R. Ayres, Northampton, UK: Edward Elgar.

Geels, F. W. 2002, Technological transition as evolutionary reconfiguration processes: a multi-level perspective and a case study, Research Policy, 31 (8/9) 1257–74.

Geels, F. W. 2004, From sectoral systems of innovation to socio-technical systems: Insights about dynamics and change from sociology and institutional theory, Research Policy, 33 (6–7) 897–920.

Greacen, C. and Footner, J. 2006, Decentralized Thai Power: Towards a sustainable energy system, Greenpeace South East Asia.

Gullev, L. 2005, District heating integrated in industrial symbiosis, in Danish Board of District Heating, DBDH, (1) 2005.

Hansen, O. E., Søndergård, B., Meredith, S. 2002, Environmental innovation in Small and Medium Sized Enterprises, Technology Analysis & Strategic Management, 14 (1) 37–55.

Hansen, O.E., Søndergård B., Stærdahl J., 2010, Sustainable Transition of Socio-technical Systems in a Governace Perspective, in Nielsen K.A., Elling B., Figueroa, M., Jelsøe E. (eds), A New Agenda for Sustainability, Ashgate Studies in Environmental Policy and Practice 91–114

Holm, J. and Stauning, I. 2002, Ecological Modernisation and 'Our Daily Bread' – Variations in the Transition of the Food Sector, Journal of Transdisciplinary Environmental Studies (TES), 1 (1) 1–13.

Holm, J., Søndergård B., Hansen O.E. 2010 , Design and sustainable transition, in Simonsen, J, Bærenholdt, J.O., Büscher M, Scheuer J.D. (eds), Design Research – Synergies from interdisciplinary perspectives, Routledge, London 123–137

International Energy Agency, IEA, 2007, World Energy Outlook 2007 (China & India).

Jacobsen, N. B., Industrial Symbiosis in Kalundborg, Denmark: A Quantitative assessment of economic and environmental aspects, Journal of Industrial Ecology, 10 (1–2) 239–253.

Jacobsen, N. B. and Anderberg, S. 2004, Understanding the evolution of industrial symbiotic networks – The case of Kalundborg. In Economics of industrial ecology: Materials, structural change and spatial scales, edited by J. Van den Bergh and M. Janssen, Cambridge, MA: MIT Press.

Kalundborg Centre for Industrial Symbiosis, 2010, Homepage info at: http://www. symbiosis.dk/ [assessed the 20th of Marts 2010].

Kemp, R. and Loorbach, D. 2006, Transition Management: a Reflexive governance approach, in Voss J. P., Bauknecht, D., Kemp, R. (eds.), Reflexive Governance for Sustainable Development, Edward Elgar, Chelterham, 103–130.

Kemp, R. and Rotmans, J. 2001, The management of the co-evolution of technical, environmental and social systems, International Conference towards Environmental Innovation Systems, Garmisch-Partenkirchen, September.

Lybæk, R. 2004, Guideline for implementing co-generation based on biomass waste from Thai industries, Ph.D thesis, Roskilde University, 2004. Download at: http://rudar. ruc.dk/handle/ 1800/ 1608.

Lybæk, R. 2008, Discovering market opportunities for future CDM projects in Asia based on biomass heat and power production and supply of district heating, Journal of Energy for Sustainable Development, 12 (2) 34–48.

Lybæk, R. 2009a, Enhancing the Sustainable Development Contribution of CDM in Asia, post doc. research, VDM Verlag, Dr. Müller, Saarbrücken, Germany, (2009), ISBN: 978-3-639-18908-7.

Lybæk, R. and Jacobsen, N. B. 2009b, The Clean Development Mechanism and the Principles of Industrial Ecology – Exploring Interconnections and Mutual Opportunities, Journal of Progress in Industrial Ecology, 6 (1) 11–28.

Lybæk, R. and Andersen, J. 2010, Enhancing the sustainable development contribution of future CDM projects in Asia, Journal of Progress in Industrial Ecology, 7 (1) 6–34.

Navanakorn Industrial Promotion Zone, 2008, Homepage info at: http://www. navanakorn.co.th/ index_r1_eng.html [assessed the 2nd of February 2008].

Opatvachirakul, S. 2009, Energy Policy and Planning Office, EPPO, Interview, Bangkok, Thailand the 18th of Marts 2009.

Pollution Control Department, PDC, 2010, Homepage info at: www.pcd.go.th/ index Eng.cfm [assessed the 12th of February 2010].

Pongvutitham, A., 2010, Drive to boost SMEs contribution to GDP, The Nation, at: http://www.nationmultimedia.com/2010/01/18/business/business_30120518. php [assessed the 23rd of May 2011].

Rip, A. and Kemp, R. 1998, Technological change, in Rayner S. and E.L. Malone (eds), Human choice and Climate Change, Battelle Press, Columbus OH, Vol. 2 327–399.

Remmen, A. 2006, Integrated product policy in Denmark – New patterns of environmental governance, in Scheer D. and Rubik F. (eds.), Governance of integrated policy – In search of sustainable production and consumption, Greenleaf Publishing, Sheffield, 103–125.

Smith, A. 2003, Transforming technological regimes for sustainable development: a role for alternative technology niches?, Science and public policy. April, 127–135.

Schumpeter, J. A. 1975, Capitalism, Socialism and Democracy, New York: Harper, 1975 (original 1942).

Schumpeter, J. A. 1975, Capitalism, Socialism and Democracy, New York: Harper, 1975 (original 1942).

Stærdahl, J., Søndergård, B., Hansen O. E. 2006, Sustainable transition of socio-technological systems: How can Governance Network Research and Transition Theory contribute to the transition to biofuel for transportation?, Paper for the Conference: Democratic Network Governance in Europe – Past and Future Research, Roskilde University, 2–3 November.

Søndergård, B., Hansen, O. E. , Holm, J. 2004, Ecological modernization and institutional transformations in the Danish textile industry, Journal of Cleaner Technology, Vol. 12 337–352.

Voss, J. P., Truffer, B., Konrad, K. 2006, Sustainability foresight: Reflexive governance in the transition of governance systems, in Voss, J. P., Bauknecht, D., Kemp, R. (eds.), Reflexive Governance for Sustainable Development, Edward Elgar, Chelterham, 162–188.

Weaver, P., Jansen, L., Grootveld, G., Vergrath, P. 2000, Sustainable technology development, Greenleaf Publishing, Sheffield.

Chapter 2.4

Opportunities and Challenges for Innovation in the Design of Low-carbon Energy Technologies – a Case Study of the Lighting Sector

Araceli Bjarklev, Kent Laursen, Jan Andersen and Tyge Kjær

Introduction

The total electricity consumption across the EU-27 countries showed an absolute increase of 28.7% between the years 1990 and 2005. The average electricity use per capita in the EU-27 is almost 2.5 times the global average. Total world consumption of energy is projected to increase by 44% from 2006 to 2030 (EIA 2010). These tendencies for growing electricity consumption still take place despite the emergence of countless numbers of energy saving devices. At the same time, the production and use of electronic devices is growing rapidly. The amount of resources required to fuel this consumption and production at a global scale is leaving a huge ecological footprint, and the EU contributes to this significantly. The decision of the European Commission to ban incandescent lighting has opened a huge debate on whether we have the technology in place to replace the Edison bulb in terms of price and current consumer demands and habits.

One of the main reasons for banning incandescent lighting is that the world cannot continue to increase its consumption of electricity because the levels of CO_2 are already surpassing the limits of a sustainable ecological footprint. When referring to a sustainable footprint, we are referring to the available energy sources (today primarily fossil fuels), the impact on the resources necessary to absorb their emissions, and to the impact of these materials and chemicals on the environment (Wackernagel 2005).

Although there is broad consensus that one of the solutions to the current environmental challenge will be based on the use of low-carbon technologies, and even though there is substantial potential to adopt more sustainable design and innovation, there are several elements that need to be taken into account to achieve efficient reductions of energy and CO_2 emissions. At the same time, and specially concerning lighting devices, to ensure their strategic implementation it will be necessary to design a product which is attractive to the consumer in terms of price, level of service and aesthetic demands. Several studies (Sandhal et al.

2006, Mert et al. 2008 and Lefèvre et al. 2006) have pointed out that, besides technological characteristics, some of the main historical barriers to the successful implementation of energy saving devices have been price competitivity with current market devices and the aesthetic design. In the case of lighting devices, this refers to the quality of light, including colour, intensity and distribution, which are related to the design of both the lamp and luminaire. This chapter takes the example of the Danish office lighting sector as a case study and discusses the question: What are the main opportunities and challenges for the design and innovation of low-carbon lighting technologies?

To answer this question, we use a systemic approach including environmental, economic, energy and political issues using relevant concepts from the Ecological Footprint, concepts and tools from Life Cycle Assessment (LCA), relevant elements from eco-efficiency and diffusion of innovation theoretical frameworks. Often, systemic approaches tend to be driven by completely rational models. However, our main contribution is to consider a more holistic approach integrating socio-psychological aspects such as consumers' perceptions (aesthetic disposition, habits and different light tastes and needs) to plan a more strategic implementation or to increase the opportunities to bring innovation in this sector to the market. Our empirical material is based on iterative interviews with relevant actors and experts within the Danish lighting sector and a preliminary consumer (user) accept test that included qualitative interviews and a quantitative survey. Furthermore, as a central part, with the cooperation of a trans-disciplinary team, our research included the design, fabrication and test of a new hybrid illumination system based on optical fibres and Light Emitting Diodes (LEDs) (Elforsk Projekt no. 341–043) .

Methods

We have as our standpoint a problem-oriented design and innovation approach, which allows for four different possibilities:

1. We take a holistic and systemic approach that ensures relevant innovation and avoids partial and unrealistic solutions.
2. We pursue a combination of technological and social solutions, where all aspects of the production chain are considered. However, we focus our approach on the most problematic stages (in our case the use phase). Even when this approach is technologically based, we further consider its societal implications so the new technology can be implemented.
3. We aim to obtain a synergy between the eco-design and eco-innovation by pursuing several benefits such as the reduction of climate impact and the reduction of the use of energy based on fossil fuels, in combination with socio-economic aspects such as local business development.

4. We pursue a strategic implementation of the innovations by using relevant elements from the Stages in the Innovation Decision Process Model from the Diffusion of Innovation theory (Rogers 2003).

Our conceptualisation of eco-design and eco-innovation is a combination of product-service life cycle and the innovation life cycle to find a more strategic implementation, as shown in Figure 2.4.1, combining the relevant elements of the following approaches:

- Using Eco-design (ISO 1462) or design for the environment, taking as a standpoint the life cycle or the production chain as a precondition for the new design.
- Using a System perspective, the design takes into consideration the intersection of combined product chains to pursue an optimal advantage or effect, in other words, switching the focus from only being single product oriented to also being service oriented. For example, instead of focusing at the lamp as a final product, we focus at the kind of service that has to be provided, in this case the service of illumination. We use a product-service system approach (UNEP 2002), which is defined as the result of an innovation strategy, shifting the business focus from the design and selling of one physical product only, to selling a system of products to deliver an integrated service. Consequently, the lamp passes from being a final product to being one of the inputs necessary to provide the service of illumination. In this way, besides considering the effects of the lamp we are also able to evaluate alternatives for other inputs, such as electricity based on fossil fuels, and find ways to improve the whole product-service system.
- Using the design concept from Eco-innovation, derived from the EU's program (COM 2009), which seeks a combination of the two previous concepts.
- Strategic and consumer-oriented innovation is pursued by including user-practice in the process of design and innovation by letting consumers test our prototypes at different stages of the innovation-decision process. By doing this we achieve a more effective dialogue with and knowledge of potential consumers. We also provide more information about the comparative advantages of our new product thereby speeding up the rate of diffusion and, at the same time, achieving a more effective consumer-oriented design.

The product-service system analysis was supported by iterative interviews with relevant actors and experts within the Danish lighting sector. We further investigated how potential users within the office lighting sector perceived our design by conducting qualitative interviews to identify the most important functions and qualitative parameters that an ideal lighting system should have in order to be accepted by the consumer. Ten face-to-face random in-depth interviews with university staff were carried out. The most important factors from the qualitative interviews were then used to design a consumer quantitative test allowing volunteers from university staff to test our system. Thirty-five persons answered the questionnaire. The whole process involved a total of 45 persons. We must stress that the test was designed to explore further considerations to support our design decisions rather than as statistical scientific research. Thus, the results of the test have been used to further re-design and develop new hybrid lighting systems from which members of our team wrote two patent applications.

The total life cycle cost of the three systems was calculated using DEEP's LCC tool (http://deep.iclei-europe.org). Because of its simplicity, this tool was chosen to assess the total energy consumption both for the initial and for future costs.

Tackling the Climate and Energy Challenge

Rising electricity consumption in the lighting sector is very problematic as the production of electricity in most countries is mainly based on fossil fuels, therefore the production of CO_2 will increase as consumption grows. Furthermore, the increasing urbanisation process in the world demands more lighting services. Levels of lighting services in industrialised countries are still too low compared with consumers' current standards and demands. Thus, the main challenge is to reduce CO_2 emissions, taking into account that consumers will not accept a reduction in the quality of service. This might explain why even when there are energy-saving alternative bulbs available, the electricity consumption due to lighting is still growing. On the one hand, it is necessary to design technologies that maintain or even improve the service of lighting using less energy, or using less energy based on fossil fuels. On the other hand, and perhaps even more importantly, is the need to ensure consumer and user acceptance, so the new technologies can be effectively implemented. By effective implementation we mean completing the innovation cycle, from designing a new product or service to ensuring its way to market.

The best lighting technology available today for the office sector, according to the European Assessment for Energy-using Products (EuP), is the tri-phosphate fluorescent lamp using electronic ballast. However, even implementing this technology (which is considered the best available) the consumption of electricity compared to 1990, will require *25%* more energy and produce *66%* more emissions of persistent organic pollutants in 2020, while emitting almost *30%* more CO_2 than in 1990 (Van Tichelen, et al. 2007). Note that the different percentage values are due to a changed relative distribution between the sources of power used in electricity production.

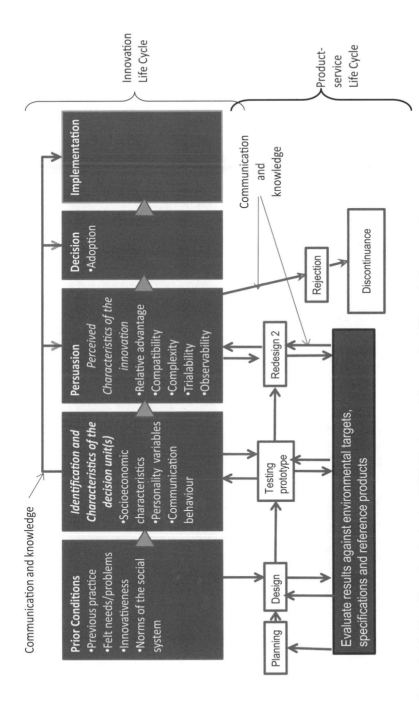

Figure 2.4.1 User Practice Inclusive ECO-Design and Innovation life cycle framework

Environmental Impacts Within the Life Cycle

In order to find an environmental solution to the energy consumption problem within this sector, we needed to determine in which phase of the lighting technology life cycle the most significant environmental impact occurred, and which processes should be improved to reduce it.

Thus, we chose to analyse this problem from a life cycle perspective. To achieve this objective, we made use of latest life cycle assessments that had detected and measured the main impacts caused by incandescent, fluorescent and LED lamps in Europe (Van Tichelen, et al., 2007 and OSRAM 2009). These assessments were important to identify the stages where a new design was needed focusing in the life cycle single approaches suggested by Coles (2002). The biggest environmental impacts were shown to be in the use phase. This helped us to focus on the specific efforts and eco-design demands in this specific stage (see Figure 2.4.2).

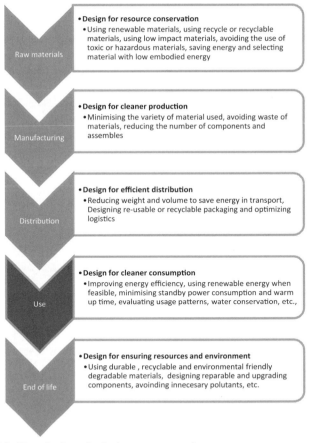

Figure 2.4.2 Eco-design single issue approaches

Improvement Potential Within the Product-service System

One of the main tendencies in the lighting industry has been to improve the energy efficiency of the lamp/bulb. The most representative technologies are linear fluorescent lamps (LFLs), and more recently, light emitting diodes (LEDs). On looking at the product-service system we realised that electricity consumption in the use phase of the lighting sector is actually causing the biggest environmental impact. Consequently, it became our objective to analyse how to reduce energy consumption and losses throughout the product-service system (see Figure 2.4.3) and to find alternative substitutes for electricity based on fossil fuels in order to deliver a more sustainable lighting service.

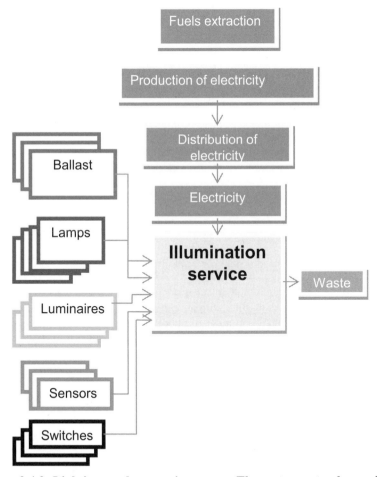

Figure 2.4.3 Lighting product-service system. The system not only considers the lighting source but also all the other necessary inputs to deliver the whole lighting service

Efficiency Challenges

When considering a lighting product-service system it is necessary to consider efficiency from three different perspectives. In general, efficiency is considered as the amount of input as a proportion of output. We focused on electricity, as it is one of the main inputs. Even if the system can convert 100% of electricity to light, it does not mean that all the light can be seen by the human eye. So, it is important to consider different types of efficiency (see Figure 2.4.4 and Box 1). Schubert defines three different kinds of efficiency: *internal quantum efficiency*, *external quantum efficiency,* and *power efficiency*.

Box 1

Internal quantum efficiency refers mainly to the input of electrons compared to the number of photons produced in a given area.

External quantum efficiency measures how effective the process is, for example comparing the number of electrons that come into the system and the number of photons that one actually gets out of the device.

Power efficiency tells us how much electrical power was needed to produce a given amount of optical power.

Luminous efficacy measures how efficient a given source is to produce light that the human eye can see.

(Schubert 2007)

However, although we can produce a certain number of photons in a given area, it is uncertain whether we can get them all out of the device. To know how much the output really is we need to look at the *external quantum efficiency* which measures how effective the process is, for example comparing the number of electrons that come into the system and the number of photons that one actually gets out of the device. The third category, *power efficiency*, tells us how much electrical power was needed to produce a given amount of optical power. This is useful in order to establish how much electrical power was lost in the process and how much electrical power one needs to produce one photon (Schubert 2007).

Although it is important to know how much optical power is available, it is also important to know how much the human eye can see from that light, since the human eye can only perceive certain frequencies of light. The *luminux flux* tells us how much visible light the human eye can really see. Thus, in order to measure how efficient a given source (the light source) is at producing light that the human eye can see, we need to measure it in terms of *luminous efficacy* (Schubert 2007). Traditionally, this sector has focused on analysing how much electricity we use per photon (external quantum efficiency) with a high luminous efficacy, so users can see the light or have a better lighting service (red circle, see Figure 2.4.4). This comparison is what Schubert (2007) calls *luminous efficacy.*

However, tendencies for continuing development indicate that the floor space area of an office will increase, and therefore so will the consumption of lighting devices. Increasing office space will result in an increased consumption of energy. Thus, from a technological point of view, the most relevant alternatives will be those with the potential to further increase luminous efficacy and which at the same time can reduce the consumption of electricity based on fossil fuel.

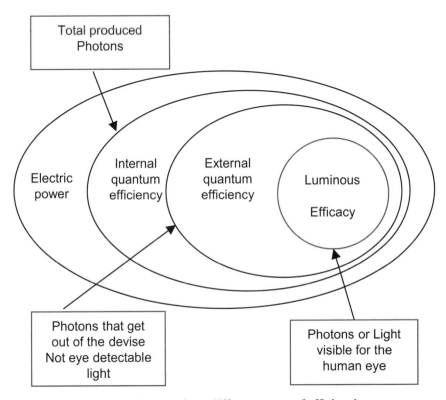

Figure 2.4.4 Illumination service – different types of efficiencies

During our research we established that one of the main technological limitations is set by the physical limits of increasing luminous efficacy with currently available technologies (fluorescent lights and LEDs). However, we saw the possibility of including technologies that could make use of renewable energy sources such as daylight in a more useful way than windows do at present. This was the result of looking at current consumption patterns or habits in the office lighting sector in Denmark. By doing this, we established that 38% of electricity used for lighting is used between 7 am and 5 pm (Elsparefonden 2010), a time when sunlight is available.

This finding suggested a variety of further technological possibilities, in particular those relating to direct light and/or solar energy. We saw the possibility of using hybrid lighting as one of the potential innovations. From the great variety of daylight transport systems, we chose to focus on fibre optical solar lighting systems. We initially wanted to test a hybrid system and compared the energy saving with a lighting system based on linear fluorescent lamps such as the one that is recommended as the best available technology by the EU Eco-design Preparatory Studies for office lighting, and the best available technology within LEDs. However, though we made an extensive search for this kind of product, there were few companies on the market advertising such systems. Moreover, when we contacted them, only Sunlight Direct Inc. responded, and told us that we could not have this product during the year of our test, 2009–2010. Not finding this product on the market, we proceeded to design and make our own prototype. This was done with the cooperation of the Technical University of Denmark (DTU Fotonik), designers from Kolding School of Design, and the assistance of IBSEN, an electrical installation contractor. To design our prototype we used a solar light transport system based on optical fibre from the company PARANS (note that the PARANS systems were not hybrid at the time that we conducted this project) and therefore we designed a system that could integrate electrical illumination (artificial light) in combination with natural light (daylight transported through fibres). For these purposes, an on/off switch system was designed and special hybrid lighting luminaires were also designed.

Although one of the main concerns in general is the question of what should replace incandescent lighting, especially when one looks at the domestic sector, the question has to be differently formulated for the office sector. In the office sector, there will be other challenges such as replacing the inefficient linear fluorescent lamps (LFLs). The problem here is that the limitation in relation to mercury content will set a very strong barrier to the further increase of luminous efficacy of LFLs and this will cause problems in fulfilling the Danish standard of providing 500 lumen per m^2 in working areas (DS 700:2005). Complying with the required Danish standard will mean a severe increase in electrical consumption and thereby potential disposal of mercury into the environment.

Recently, LED technology has been seen as the favourite to replace even LFLs. Although the EU preparatory studies for the eco-design directives for office lighting (Van Tichelen, et al. 2007) considered that LEDs were not yet ready due to the colour of the light emitted, lumens per watt and price. Further developments over the last three years have made it possible to consider this technology as a viable solution for the future in relation to colour rendering and temperature of the light. In relation to price, findings show that LED prices are still high compared to conventional devices.

Opportunities

It is important to bear in mind that the EU preparatory studies for the eco-design directives on office lighting (Van Tichelen, et al. 2007) pointed out that some of the main improvements in this sector can be achieved by including sensors, improving the maintenance level of the lighting system and, most importantly, by including natural daylight. These recommendations are important because the studies consider the whole lighting service and not only the lamp's life cycle. Therefore further opportunities beyond those relating to the lamps may exist.

These improvements consist not only of increasing the efficiency of the lamps but also of making the systems easier to maintain and clean, and of including sensors both for dimming the light, when there is more daylight available, and for detecting movement. By doing this, savings of between 50% and 80% of the total consumption of electricity may be achieved (Van Tichelen, et al. 2007 and Verhaar 2010).

Considering daylight, we can envisage further opportunities. One obvious possibility is the use of windows. Consequently some of the current responses to these challenges emphasise the use of more windows. However, recent studies show that when more windows are included the increased load on the heating or cooling systems increases the overall energy demand of the building (Asdrubali, et al. 2010).

In practice, the use of windows can result in additional use of electricity for lighting because the users close the blinds and turn on electrical light (see Figure 2.4.5). This picture shows an office at a research institution and was taken when there was full sunshine outdoors during summer 2009. Notice that the artificial light is on.

Figure 2.4.5 Consumers' habits using windows for lighting purposes

The impact of constructing new buildings with more windows is relatively small as the number of new buildings represents a very small percentage of the whole building mass in Denmark. Although it is very important to construct new buildings in a more sustainable way, one of the biggest challenges in this sector is how to include more daylight in the existing private and public office buildings within the service sector without causing glare or increasing the heating or cooling loads inside the buildings, thereby worsening the environmental working conditions.

Therefore, we focused on light transport systems that were flexible and easy to install within existing typical buildings. These features were found in solar optical systems. Although light transport systems are already known, little attention has been paid to them. However, they provide interesting opportunities for future lighting system applications. Thus, in cooperation with a trans-disciplinary team, we designed a hybrid lighting system (see Figure 2.4.6). This was based on a commercial solar transport system from PARANS using commercial LEDs. Financial support for this work was obtained from the Danish Elforsk program.

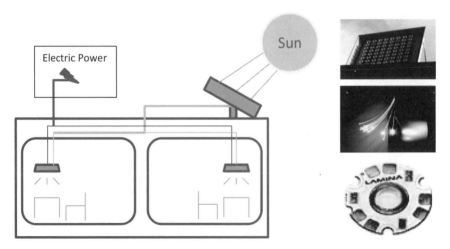

**Figure 2.4.6 Hybrid lighting system, at the right solar collector, optical
fibres and LEDs**

The main idea was to produce a hybrid lighting system using electric power and direct sunlight to provide a high quality and constantly reliable lighting service. Hybrid lighting systems are a combination of four technologies: collecting natural light, generating artificial light, transporting and distributing light to where it is needed, and controlling the amounts of both natural and artificial light continuously during usage. The overall idea is not new. The US Department of Energy (DOE), in collaboration with the Oak Ridge National Laboratory, has published a large number of articles pointing out the advantages of hybrid solar fibre optical systems in terms of energy savings. However, when we started this project in 2009, it was not possible to purchase a hybrid lighting system based on fibre optic technology.

It was only possible to buy solar systems based on fibre optic technologies. When both systems are installed without an automatic switch they only add an extra cost and no energy savings are obtained.

The main principles behind designing a hybrid luminaire are:

- In order to harvest energy savings, electrical power is switched off automatically when the sun is shining with a sufficiently high intensity to provide a pleasant indoor light intensity.
- Electrical power is switched on automatically when it is dark or cloudy.
- The system can use a standard electricity supply either from renewable or non-renewable networks (smart grid) working even when it is overcast.

Designing and testing this hybrid system led us to some very interesting results. First of all, we calculated that considering all the inputs of energy and outputs in lumens for our three comparison cases - LFL, LEDs and LEDs/solar optical transport system (hybrid system) - the energy savings, and thereby the CO_2 savings, to be 59% as they are proportional to the energy savings as shown in Figure 2.4.7.

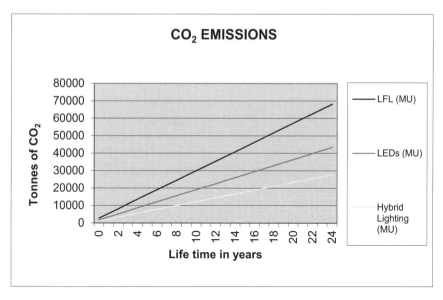

Figure 2.4.7 CO_2 emissions for the three options for a Multiple User office (MU) using the DEEP's LCC tool (http://deep.iclei-europe.org)

Consumer Acceptance and Price

According to Rogers (2003) the diffusion of innovation, or the rate at which the consumers will accept or reject an innovation, will depend of the communication channels used to communicate the benefits of the product or service. Although

Rogers describes several types of communication channels, the interpersonal ones are of interest to us. He argues that often the most successful methods for the diffusion of innovation take place when an innovation has been conveyed to people by other individuals like themselves who have already adopted that innovation. Therefore, identifying the main stakeholders and consumers/users that will adopt this innovation is of central importance. As with any kind of environmental innovation, consumers' acceptance and price are decisive for any technology to be implemented (Kjær and Andersen 1993). According to Rogers (2003), not all innovations are equivalent in terms of analysis, and therefore it is important to identify the characteristics of innovation as perceived by individuals, as well as the type of individuals. As we have stated earlier, the main problem in relation to CO_2 emissions is in the trade and service sector therefore we focused on these individuals, identifying the main stakeholders and their perceived needs.

Identification of Decision Makers

One needs to identify who decides what lighting system should be bought to satisfy the demand for public buildings. When it comes to office lighting there are several stakeholders who influence the decision-making process about which type of technology should be adopted. In order to establish whose habits and established traditions had to be considered we needed to identify the role of these stakeholders clearly. For this specific study we identified: 1) regulators, 2) service providers, such as consultancy firms, lighting designers, architects, engineers and urban developers, 3) consumers, such as those who pay for the service and for whom the price will be important, and 4) users, such as employees who use the service but do not necessarily pay for it. For this last group the relative advantages of the innovation will be more important than the price.

Price

In order to assess the cost, we used DEEP's LCC tool (http://deep.iclei-europe. org). This tool was chosen for its simplicity to assess the total energy consumption both for the initial and future cost. It takes into account the costs, such as materials and installation, and evaluates them in terms of Net Present Value.

To calculate the Net Present Value we assumed a current discount rate of 3% (Denmark National Bank, 25 March 2010, ref. 2010-14E), an 'on-peak electricity tariff price escalator' of 3% as an average value for the different energy sources (Bolt, 2009) and a carbon factor of 0.801kg CO_2/kWh (DONG Energy, 2008). We further considered that the lifetime for the hybrid system will be 25 years, as the LEDs will be used less. The emission factor reported by the EU Commission is 0.041kg CO_2/kWh. However, we do not consider this value, as it is representative of an average of all kinds of energy sources (renewable and non-renewable).

We chose to consider 0.801kg CO2/kWh, since it is more representative for the marginal energy produced by non-renewable sources. The comparison of the base cases is shown in Figure 2.4.8.

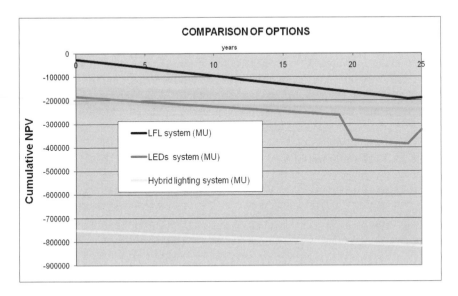

Figure 2.4.8 Comparison of options in terms of Net Present Value

As can be seen from Figure 2.4.8, the main barrier for this design is the price of the system tested. The life cycle cost is almost eight times higher than the LFL system and four times higher than the LED system. Thus, even when the hybrid system is the best option for lower CO_2 emissions, the price will limit its implementation. Taking the whole life cycle of the hybrid systems into account, we assessed that the contributing factors to this barrier were complexity of the electronic system to drive the lenses in the solar collector system, the quantity of material used and the installation costs. Therefore, in order to improve the design of the solar collector system we suggested going back to the Eco-design single-issue approaches (Figure 2.4.2) and investigating the materials and manufacturing stages in this specific product. We recommend as well further research and development in this direction.

Required Innovation Characteristics

This requires looking at subjective preferences or properties in relation to function. In the case of values ascribed to lighting during the development of this research, we have learned from dialogue with key stakeholders, consumers and users that temperature of light (colour), spectral composition and noise (non-oscillating light), as well as aesthetic and prestige values are important characteristics that will cause Danish consumers and users to accept or reject an innovation. When looking at the material dimension, as a base of comparison, practitioners of the LCA suggest considering the service provided by the product as a functional unit (see Table 2.4.1). The functional unit, according to Wenzel and Caspersen, (1999) should be defined considering both the obligatory properties (tangible dimensions) and the positioning properties (intangible dimensions) of a product or a service.

The aim is to quantify the product or service with respect to volume and time, looking for potential improvements, in this case the lamp. Shostack (1977) suggested some time ago that in order to see possibilities that can reach the market, one should pass from being product-oriented to being service-oriented. This requires being able to see the tangible dimension (materials) but also to see the intangible dimension (characteristics of the service).

Table 2.4.1   Functional unit

Obligatory properties	Positioning properties
Supplying 500 lux to a working area in a typical Danish office	– Non-blinding effect – Avoiding overheating – Optimal spectral composition of light for a working space (colour and warmth of light) – Aesthetic components of the room – Reducing CO_2 emissions – Indication of 'green responsibility'

Although LCA practitioners recognise the importance of the intangible dimension, the disadvantage of the LCA approach is that in practice the positioning properties are always defined by the designer or engineer (either working individually or in a team). Taking into account the stages model of the innovation-decision model we integrate users' practice and opinions into the re-design or re-invention process, as Rogers named it (Rogers, 2003). Rogers' point is that innovations are not invariant. On the contrary, they usually change as they are diffused (Rogers, 2003). The advantages of integrating this approach compared to that of the LCA is that including this approach helps to achieve a more effective consumer-oriented design in a very early stage of the design thereby improving the changes for a more effective implementation.

The importance of including users practice and opinion can be appreciated comparing Table 2.4.1 and Table 2.4.2. Table 2.4.1 shows the consideration that we as a design group made at the beginning of this project. Table 2.4.2 was completed after the user quantitative and qualitative test. One of the main issues when architects and lighting designers in particular discuss 'quality of light' is that the light should be warm and therefore as close to the light from incandescent lamps as possible. During the development of this study we met some architects who expressed their disappointment with the new regulations, since in their opinion; there was no satisfactory substitute for the incandescent lamp. To our surprise, when we asked users in the office sector, they stated that if the light were soft and warm they would fall asleep in their offices.

Table 2.4.2   Functional unit including users' practice and opinions

Obligatory properties	Positioning properties
Supplying 500 lux to a working area in a typical Danish office	– It should be possible to read and write comfortably – It should be possible to work at a computer without being affected by reflected light – It should be possible to see peoples' faces while holding meetings – The light has to be as close to daylight as possible (indirect sun light) both in colour and intensity – Avoiding overheating – Direction of lighting should be at a comfortable angle (no blinding) – The lighting should be aesthetically pleasing – Ability to be flexible (regulation of intensity) – Reduce CO_2 emissions – It should not contain mercury – Indication of daylight usage ('green responsibility')

Designing a hybrid lamp

The contribution of our project partners from Kolding School of Design consisted of the design and construction of two prototypes of hybrid luminaires. One of them is an integrated, ready-to-install end solution (see Figure 2.4.9 and 2.4.10). The other luminaire is an adaptable solution designed with the intention of giving consumers the option of bulb replacement in existing luminaires. This option offers the possibility of conserving classical luminaires that consumers value highly, for example famous Danish designed luminaires (see Figure 2.4.11). Light emitting diodes (LEDs) were selected in the design of both luminaires because these kinds of devices do not contain mercury and have a longer life span. From a technological point of view, LEDs are more flexible for use together with on/off switches when interacting with solar lighting, achieving real energy savings. They have the potential to become more effective in the future. Their intensity and colour of light is flexible, so they can be designed to fulfil consumer expectations. From an aesthetic point of view, their size and flexibility allows for many new forms and expressions. The lamps were designed with an automatic intelligent on/off switch that allows the change from natural light to artificial light by making use of two fibre optic bundles to provide the direct solar light. In this way, even when it is overcast, the lamps can deliver a constant and reliable light source at all times of the day. Besides these two lamps, design students from Kolding School of Design were invited to participate to learn more about this technology as they will be key stakeholders in the lighting sector in the future.

Armature suspension

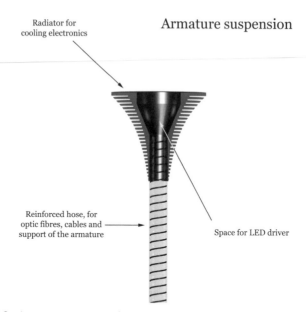

Radiator for
cooling electronics

Reinforced hose, for
optic fibres, cables and
support of the armature

Space for LED driver

Figure 2.4.9 Armature suspension

Hybrid armature

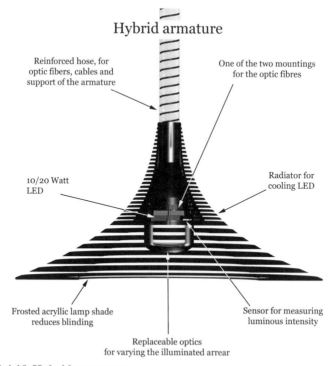

Reinforced hose, for
optic fibers, cables and
support of the armature

One of the two mountings
for the optic fibres

10/20 Watt
LED

Radiator for
cooling LED

Frosted acryllic lamp shade
reduces blinding

Sensor for measuring
luminous intensity

Replaceable optics
for varying the illuminated arrear

Figure 2.4.10 Hybrid armature

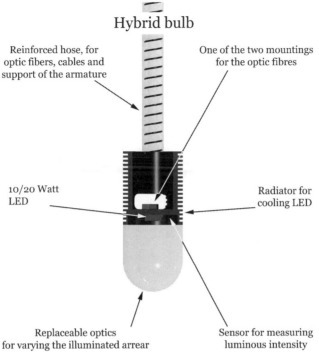

Hybrid bulb

Reinforced hose, for optic fibers, cables and support of the armature

One of the two mountings for the optic fibres

10/20 Watt LED

Radiator for cooling LED

Replaceable optics for varying the illuminated arrear

Sensor for measuring luminous intensity

Figures 2.4.11   Hybrid bulb

Testing our first prototype at Roskilde University in one of the main buildings

We conducted a user focus group. We installed the fibre optic solar collector system from PARANS and the hybrid luminaires that were designed during this project including those designed by the students. On the basis of our explorative qualitative interviews with user focus group, we formulated a questionnaire. The questionnaire worked as the document test for reading and writing tasks, and we provided a laptop computer for the use of respondents. The test group was composed of staff and student volunteers at Roskilde University who agreed to test the system. The system was set up in one of the rooms near the central cafeteria of Roskilde University (see Figure 2.4.12, 2.4.13, 2.4.14 and 2.4.15).

Suitability for the Performance of Office Work

From consumer feedback we concluded that there was broad acceptance that the service fulfilled user criteria. We also realised that older people required greater light intensity in order to read. Reading and writing using a computer seemed to be an improvement, even considering that the computer has its own light source. In relation to holding meetings, the lighting provided by the hybrid system was

Figure 2.4.12   Installation of the solar collector

Figure 2.4.13   Installation of the optical fibres

Figure 2.4.14   Installation of the luminaires

Figure 2.4.15   The layout of the room

satisfactory since this activity does not necessarily require too much reading and writing. From these responses we concluded that, for general lighting purposes, the light intensity is acceptable but for work task areas, the system should provide the possibility of increasing the intensity according to the user's physiological demands.

Light Dynamics

In relation to our test, 91% of the respondents had experienced lighting variations. This occurred, when the sky was overcast and when the sun started to shine again or vice versa. Nine per cent of the respondents did not experience lighting variations. This might be because there was sunshine or an overcast sky during the entire period of the test.

There are some losses from the red part of the spectrum in relation to fibre optic lighting and this was observable in our test, since the designers used warm white LEDs. As indicated in our qualitative interviews, light dynamics were within the parameters that our focus group considered important. On further enquiry into whether they liked the light emitted by our system, the reasons given for this were that they found the light pleasant, that there is an indication of daylight, and that the light appears natural.

We learned that most people liked the light dynamics of the test system, especially those in the 20–30 and 30–40 age groups. The main reasons given were that they found it *pleasant* and *because it demonstrates the use of daylight*. This was a positive indicator as one of the positioning properties listed by the users suggested that light should have this value. From this, we can conclude that even when the light emitted by the optical fibres is a bit greenish, *it is not a big limitation* for the user to accept or like the system as it was tested. However, further product development and tests should be carried out to also reach population older than 50 years old. In order to have a fuller and broader acceptance from all types of user, the system should be as close to the natural light (daylight) spectrum as possible, considering the qualitative interviews, and this should be as important for the LED colour selection as for the light coming from the optical fibres.

Temperature, colour and quality of light

Taking into account the feedback below, our team is currently designing a series of new improvements that can better satisfy our test users in the future, as we are convinced that not only has the technology strong energy and CO_2 savings potential, but also strong consumer acceptance potential. One of the next steps is to make a more permanent installation so that not only end users, but also other decision makers can test systems like these. Further, our task also consists of improving the price in order to be competitive with current lighting systems.

Light Temperature

When the users were asked if they experienced warm or cold light, 71% of the interviewees answered that they had experienced more cold light. This might be due to the fact that the system was operating in full sunshine most of the time and the LED light took only part of the time.

Colour of Light

In relation to the colour of the light, to our surprise, 67% of the users indicated that they experienced blue light. As we mentioned before, one of the characteristics of the PARANS solar system is that the spectrum of light is dominated by the green part. However, it is interesting to see that what users perceived as cold light is related to the colour blue.

Blinding Effect

We allowed the respondents to mark more than one option in relation to the quality of light. When the respondents marked more than one option, we gave all responses the same weight. 70.9% of respondents answered that the light did not blind them. This was a positive answer. However, we need to improve the design so that all the users are satisfied.

Even Distribution of Light in the Room

It is important to remember that some of the lamps were prototypes developed in a five week project by students of Kolding School of Design and not commercial examples. These lamps were fragile, and although we asked the users not to touch them, many of the test users wanted to see the interior of the lamp, the LEDs, the optical fibres and the automatic switch. This showed that there is great excitement from consumers when using new types of lighting that emit natural light and which they are therefore not used to. Due to this, some of the lamps' electrical systems stopped working and were operating with only natural light, so when it was overcast only the remaining lamps worked and the room was only partially lit. It is not surprising that only 40.6% of the users thought that the light was distributed evenly within the room. From this we learned that our new prototypes have to be very robust to be able to satisfy users' curiosity.

Influence of the Armature Design

Both women (50%) and men (75%) stated that they had been influenced by the luminaire's design when considering their acceptance of this innovation. This means that we need to continue working on this area when we ask users to test the technology and we also need to improve the design for the final commercial product.

Conclusions

One of the main limitations to planning strategic implementation of eco-design and innovation within the clean technologies sector is the lack of broad holistic approaches. Existing approaches refer to isolated parameters such as function–cost or environmental issues such as energy efficiency and cost assessment. They also lack the inclusion and evaluation of characteristics of innovation as perceived by individuals. Not including a consumer-oriented approach might result in slow introduction to the market, or even implementation failure (rejection), which can be very expensive and increase uncertainty for potential investors or business developers. On the other hand, isolated consumer-oriented research may result in technological and environmental failures, as consumers often are not experts, nor are they aware of the many environmental challenges as they are very complex issues, and this may also result in innovation failure.

The main benefit of using our suggested framework is that at the same time as revealing the technological, environmental and cost improvements, it also leads to the possibility of innovative solutions and the opportunity of new business niches. The methodology also provides, in a simple way, a good overview of consumer needs and the opportunity to communicate the properties of a new product or service. It further provides the opportunity to receive feedback from the consumers' point of view in relation to the suggested innovation, and in this way includes the consumer in the process of redesign and diffusion from an early stage making it possible to plan the implementation more strategically.

The main limitation for hybrid optical fibre lighting systems is still their initial high cost. However, looking at the material and production processes, to redesign them could bring the price down and thereby make them more price-competitive during their life cycle. During the test phase, we have made some design suggestions. An exploratory consumer (user) test, even if not strongly scientifically based, provided us with a relevant and useful insight into the challenges and design potential of making this technology more acceptable to consumers and to see the product's potential to provide users with better quality lighting. At the same time, we foresee the potential for a great variety of aesthetic applications and, more importantly, a promising potential niche market.

References

Andersen, P. D., Borup M., and Olsen M.H. 2006. Innovation in energy technologies. In Larsen H. and Petersen L. Risø Energy Report 5 – *Renewable energy for power and transport*. Risø National Laboratory, 2006. Pp. 21–28.

Asdrubali, F., Baldinelli, G., and Baldassarri, C. 2010. Life Cycle Assessment of buildings and electric energy consumptions. pp 253–260 in *Proceedings of CIE 2010 "Lighting Quality and Energy Efficiency"* 14–17 March 2010. Vienna, Austria. ISBN 978 3 901906 83 1

Bolt, J. 2009. *Fremtidige priser på biomasse til energiformål.* Available at: http://www.ens.dk/daDK/Info/TalOgKort/Fremskrivninger/modeller/Documents/Biomassepriser20090120.pdf .Accessed the 13-03-10

Coles, R. 2002. Ecodesign tools section 3. In *The Sustainability Pack, draft*; ITDG (international Technology Development Group), UK 2002

COM. 2009. 519 final. Investing in the development of Low Carbon Technologies. *(SET-Plan).* Available at: http://ec.europa.eu/energy/technology/set_plan/doc/2009_comm_investing_development_low_carbon_technologies_en.pdf. Accessed the 20 may 2011

DONG Energy. 2008. ASNÆSVÆRKET- *Grønt Regnskab 2008.* Available at: http://www.dongenergy.com/SiteCollectionDocuments/NEW%20Corporate/PDF/Groenne_regnskaber_2008/Groent_regnskab_2008_Asnaesvaerket.pdf. Accessed the 18-08-10

DS 700: 2005. Kunstig belysning i arbejdslokaler. Danish Standards association

EIA. 2009. *World Energy Outlook 2009.* Executive Summary. International Energy Agency. Pp 17. Available at: http://www.worldenergyoutlook.org/docs/weo2009/WEO2009_es_english.pdf. Accessed the 19th of March 2011.

EIA. 2010. *International Energy Outlook 2010.* Available at: http://www.eia.doe.gov/oiaf/ieo/highlights.html. Accessed the 20-10-2011

Elforsk Projekt nr. 341–43 (2010). *Hybrid Fiber belysning rattet mod en mindre økologiske fodaftryk.* Available at: http://www.elforsk.dk/doks/341-043/341-043_rapport%20.pdf.

Elsparefonden statistics. 2010. Available at: http://application.sparel.dk/ElWebUI/El/index.aspx. Accessed the 20-05-09

ISO 14062. ISO/PDTR 14062 working draft: 2001-10-30. *Environmental management-integrating aspects into product design and development.* International Organization for Standardization. Geneva, Switzerland.

Kjær, T. and Andersen, J. 1993. *Implementering af Vedvarende Energykilder i Norden.* Udarbejdet for Energi och Miljögruppen under Nordiske Ministerråd. ENSPAC Institut at Roskilde University. Postboks 260 Dk-4000 Roskilde. Pp. 108

Lefèvre, N., de T'Serclaes P. and Waide P. 2006. Barriers to technology diffusion: The case of compact fluorescent lamps. International Energy Agency COM/ENV/EPOC/IEA/SLT(2006)10

Mert, W., Suscheck-Berger J. and Tritthart W. 2008. *Consumer acceptance of smart appliences*. D5.5 of WP 5 Report from smart-A project. EIE. Smart Domestic Appliances in Sustainable Energy Systems (Smart-A). Intelligent Energy Europe.

OSRAM. 2009. *Life Cycle Assessment of illuminants- a comparison of light bulbs, compact fluorescent lamps and LEds Lamps*. Executive summary, Novemeber 2009. OSRAM Opto Semiconductors GmbH Innovations Management, Regensburg, Germany. Available at: http:/os.com/osram_os/EN/About_Us/ We_shape_the_future_of_light/Our_obligation/LED_lifecycle_assessment/ OSRAM_LED_LCA_Summary_November_2009.pdf. Accessed the 23- November - 2010

Rogers, E. M. 2003. *Diffusion of Innovations*. Free Press, A Division of Simon & Schuter, Inc, New York, NY. Fifth Edition. Pp 551.

Sandahl, L.J., Gilbride, T.L, Ledbetter, M.R. and Calwel C. 2006. Compact fluorescent lighting in America: lessons learned on the way to market. Prepared for the Department of Energy. Pacific Northwest National laboratory.

Schubert, E.F. 2007. *Light Emitting diodes*. Second edition, Cambridge University Press. Pp. 422. ISBN 978-0-521-86538-8

Shostack, L. 1977. Breaking free from product marketing. American Marketing Association. *The Journal of Marketing*, Vol.41, N2 (Apr. 1997), pp. 73–80. Acceded the 25/01/2010 at: http.//www.jstor.org/stable/1250637

UNEP. 2002. *Product –Service Systems and Sustainability – Opportunities for sustainable solutions*. Pp. 31. United Nations Environment Programme , Division of technologicay Industry and Economics. Production and Consumption branch. Consulted the 19-02-2010 at: http://www.unep.fr/scp/ design/pdf/pss-imp-7.pdf

Van Tichelen, P., Jensen,B., Geerken, T., Vanden Bosh, M., Vanhoof, L., Vanhooydonck, L., and Vercalsteren, A. 2007. *Preparatory Studies for Eco-design requirements of EuPs –Final Report Lot 8: Office lighting*. Vito. April 2007. Pp. 266. Consulted several times in 2009 at: www.eup4light.net

Verhaar, H. 2010. The contributions of energy efficient lighting to accelerate renovation of buildings and cities. In *Proceedings of CIE 2010 "Lighting Quality and Energy Efficiency"* 14–17 March 2010. Vienna, Austria. ISBN 978 3 901906 83 1

Wackernagel, M. 2005. Europe 2005- *The ecological footprint*. June 2005, WWF European Policy Office Brussels Belgium. 23 pp.

Wenzel H. and Caspersen N. 2000. Product life cycle Check- a guide. Institute for Product Development, Special edition adapted for course 42372, Tech University of Denmark by Dr. Michael Hauschild, September 2000. Pp. 55

Using Sustainability Games to Elicit Moral Hypotheses from Scientists and Engineers

Evan Selinger, Thomas P. Seager, Susan Spierre and David Schwartz

Introduction

This chapter outlines a new pedagogical initiative. "An Experiential Pedagogy for Sustainability Ethics" is a newly started National Science Foundation sponsored project that aspires to challenge the received wisdom depicting scientists and engineers as people who are most likely to contribute to solving sustainability problems in two ways: i) through technical research that fuels technological innovation, and ii) by dispensing technocratic advice.[1] Such a view characterizes scientists and engineers as highly pertinent players—leaders, in fact—in sustainability conversations, so long as those conversations stay away from what many consider true moral territory—territory where technocracy is a reviled ideal, and preference is given to notions of strong democracy and enhanced participation. This demarcation implies that scientists and engineers become experts by mastering cognitive styles that either are unrelated to moral reasoning or which achieve coherence by denigrating values or processes associated with moral reasoning. While plenty of anecdotal evidence validates this stereotyping, the depiction obscures a crucial point. For the most part, ethics education has failed scientists and engineers! The more this point remains unappreciated, the more difficult it will be create the partnerships needed to bring about a sustainable future.

As we see it, a new approach to teaching sustainability ethics, one that is structured around experiences that science and engineering students experience as emotionally resonant and culturally relevant, and which revolve around skill building in the domains of communication, deliberation, and collective action, is an important step for moving beyond the negative stereotypes that arise from contingency—the contingency of ethics education following a liberal arts trajectory that is divorced from the experimental world of science and engineering training. To ensure that our views are widely accessible, we will restrict our presentation to a general overview of the motivational ideas underlying "An Experiential Approach for Sustainability Ethics." To give a concrete sense of how students who take the type of course we are designing might generate their own novel moral hypotheses, we conclude with a brief presentation of how Susan Spierre, a doctoral student in

---

1   The NSF grant number is 0932490.

Sustainability with graduate background in climate science, is applying core ideas in her newly conceived research on the moral allocation of CO_2 emissions.

Barriers to Teaching Sustainability Ethics

Two main barriers make it difficult to teach an effective sustainability ethics course to science and engineering students: identifying an adequate depiction of sustainability dilemmas that have moral components, and convincing students that even if they do not initially know it, they in fact do care about ethics. We will discuss the former issue first, and conclude this section by revisiting the latter.

With some degree of strategic exaggeration, we can say that two types of images symbolize the typical approach to engineering ethics education: images of disasters and images of near-disasters. On the disaster side, one only needs to think of exploding pictures of the Space Shuttle Challenger and Ford Pinto. These didactic exemplars convey two rules that seem obvious, when understood retrospectively: don't let the folks at NASA launch a space shuttle when it is cold; don't let auto-manufactures design cars with gas tanks in the bumper. Both images make most people cringe. They readily bring to mind the dramatic consequences that can result when individuals make bad engineering decisions.

On the positive side, one only needs to envision the Citigroup Center, one the tallest skyscrapers in New York. This image is canonized in engineering ethics books. Why? Because Citicorp allowed information from William LeMessurier, an engineer, to proactively change a seemingly acceptable, but in reality faulty approach to the problem of safety in the context of designing buildings that can bear considerable wind loads.

What happens when students take ethics classes that focus on case studies of the sort just evoked? They readily see the world as exemplifying the Manichean struggle between bad apples and whistleblowers. That is, they end up interpreting overly-simplistic, thin depictions of real case studies as evidence that justifies understanding ethical conflict through the following fictional script: Ethical dilemmas appear to arise when one-dimensional bad apples fail to exhibit the good character, intellectual foresight, and empathetic concern routinely demonstrated by whistle blowers and other conscientious professionals. Within this script, issues of context and institutional behavior become minimized—sometimes going wholly unnoticed—and students end up feeling disconnected from their objects of study. In the case of bad apples, students do not identify with the professionals they read about because they overestimate their own moral fortitude, and see the caricatures they are presented with as wicked "others" who are fundamentally dissimilar to them. Likewise, students are prone to all too easily imagining themselves as playing the heroic role of whistleblower, or else relegating ethics to the domain of unusual, heroic activity.

Given the problems just detailed, it might seem useful to introduce science and engineering students to problems in sustainability that use nuanced narratives to convey the tragic results of failing to live sustainably. Appropriate case studies would focus on a scale of analysis that differs from the one conveyed by the previous examples, society, not individual decisions-makers. With this situation in mind, Jared Diamond's *Collapse: How Societies Chose to Fail or Succeed* and Carl McDaniel and John Gowdy's *Paradise for Sale* appear especially germane. Unfortunately, while these texts and others like them are valuable scholarly resources, it is unlikely that students will identify with the stories they present, such as Nauru Island being ruined as a result of poorly managing phosphate resources and the environmentally based collapse of Easter Island. The problem of "otherness" remains, as students will perceive these cultures as so dissimilar to their own as to be little more than tales of primitive people and place. When contemporary exceptionalism becomes the dominant interpretative sensibility, good books about historical occurrences become filtered through chauvinism and dismissed as inapplicable cautionary tales. After all, our students "know better."

The problem of otherness cannot be resolved simply by gaining greater insight into the images that students perceive as authentic signs of sustainability problems. Probing deeper in this direction would reveal that their imaginations readily leap to apocalyptic scenarios: the last of the polar bears perched on top of melting ice, Japanese teens wearing Hello Kitty emblazoned breathing masks, sleek astronauts off to terraform a replacement planet, *et cetera*. Potentially more problematic is the difficulty of students envisioning a viable voice for one of the core areas of sustainability discussions, future generations: What will they be like, what values will they embody, and how will they conceive of the good life? It is too easy to think of future generations as fundamentally similar to people living in the present, or as dystopian distortions, perhaps beginning life as babies who are branded are branded from head to toe with corporate logos, and live in a time where big business has commodified everything and natural treasures have been replaced by virtual simulations.

Compounding the problem of identifying adequate cases of sustainability ethics dilemmas is the problem of teaching ethics to science and engineering students. The basic problem is not about ethics per se, but concerns the liberal arts horizon and its constitutive set of pedagogical tools. When ethics is taught from this perspective, the following deep problems usually arise:

1. Science and engineering students are placed in an environment that centers on the interpretation of text and the exchange of verbal discourse that differs significantly from the experimental settings where their major studies are conducted, and they solve or give up on problems when the material world complies with or resists their interventions. This difference is not favorably perceived. Instead, liberal arts is viewed as an arena where subjective and sometimes arbitrary judgment proves decisive.

2. Students feel disconnected from the material because they are positioned to relate to it as spectators and readers. This is a general point, one that goes beyond the problems of otherness that might be more specific to

sustainability and engineering ethics education. In other words, students are removed the objects they study because they are limited to discussing decisions that other people have made or engaging in thought experiments where they imagine how they would act were they to be placed in a particular scenario. Neither instance has the direct feel of laboratory work. In the laboratory case, the material world may have the final say, but students recognize that failure to execute a technique or procedure correctly can prove disastrous, while proper execution can be an accomplishment worthy of pride.

3. While humanities approaches to teaching ethics have come to include more attention to group work, ethics education typically is presented at the scale of the individual decision maker. In reality, engineering design proceeds as the result of deliberative, team-based processes in which ethical issues are reasoned through collectively.

These problems are aptly captured in a political cartoon that depicts a college age student returning a book to a stern looking librarian, and contains the following caption: "I need a statement that I've been here for the past 15 minutes and read David Hume's *An Enquiry Concerning Human Understanding*." By philosophical standards, Hume is considered an engaging and spirited writer who offers provocative and pertinent writings on such topics as knowledge and ethics. But from the perspective of most students, an assignment to read Hume is akin to being given nearly any assignment in an ethics class: The goal is to put in the minimal amount of effort required to make the instructor think you have done more work than you actually did.

In order to find a resonant way to teach ethics to science and engineering students, instructors also have to confront a general problem with ethics education. Although students often come into ethics classes professing to be moral relativists, they tend to possess more robust intuitions concerning justice than they initially acknowledge. The trick is to create a situation that produces moral tension— tension that forces them to critically examine how they perceive, conceptualize, and even feel about matters of justice after they themselves have been subjected to moral slights. This phenomenon is aptly captured in a Calvin and Hobbes comic that has the following structure.

Echoing a caricatured version of Machiavelli, young Calvin tells his stuffed tiger Hobbes, "I don't believe in ethics any more." He then adds, "As far as I'm concerned, the ends justify the means. Get what you can while the getting's good ... Might makes right! The winners write the history books!" He concludes the tirade by asserting, "It's a dog-eat dog world. So I'll have to do whatever I have to and let others argue about whether it is 'right' or not." Provoking moral tension, Hobbes responds by shoving Calvin head first into a pile of mud, which soaks him from head to toe. Indignant at this act of seemingly unprovoked violence, Calvin demands an explanation. Hobbes obliges, using the principle of consistency to present Calvin with a version of his own egoism: "You were in my way. Now you're not. The ends justify the means." Stung by a viscerally felt understanding of his hypocrisy, Calvin ends the strip with an ironic parting line: "I didn't mean for everyone you dolt! Just me!"

Sustainability Games

To remedy the problems detailed in the last section, "An Experiential Pedagogy for Sustainability Ethics" is guided by the premise that games which simulate basic problems in sustainability ethics are an exceptionally good medium for evoking emotionally resonant, cognitively challenging, memorable, and motivating deliberations for science and engineering students who belong to the Millennial/ Digital Generation. Members of this generation bring games with them wherever they go, storing them on portable devices like iPhones and iPads. They enjoy playing games through social networking media, such as Facebook's popular Farmville, grew up immersed in massive, multiplayer online role-playing games, such as World of Warcraft, and are accustomed to experimenting with modeling parameters in virtual worlds, such as Second Life.

Perhaps more importantly, educational games of the sort referred to as "serious games"—which is the genre our project belongs to—have the potential to overcome the ethics education problems previously discussed because they enable students to be active participants in ethical dilemmas, instead of passive spectators. Furthermore, in the gaming context, students are given opportunities to experiment with different strategies, and therein experience themselves as being embedded within an ethics laboratory that closely resembles the learning environments of their science and engineering classes. This horizon is one that they are emotionally invested in, and that emotional investment can pay moral dividends, especially when game play instills a felt sense of responsibility: It can be a powerful experience for students to commit to courses of action that affect how well their classmates do, especially if, as we intend to stipulate, game play counts for part of the student's course grade, and disappointed students convey their moral indignation; it also can be a powerful experience for students to have their fate determined by the actions of their classmates—an experience akin to the one that led Calvin to reconsider his initial moral relativism. Finally, game play can avoid the problem of otherness. When students judge strategies of their own experimental design and the experimental design of their classmates, it becomes hard to reduce moral conduct to the actions of one-dimensional bad apples and whistle blowers, especially when, *as must be the case* in games that simulate sustainability ethics problems, game play occurs at a social level where collective deliberation is functionalized as a fundamental component.

The games that we have in mind are ones that capture salient issues related to the ethical management of environmental externalities (rooted in situations addressed by the Coase Theorem and the Tragedy of the Commons), the ethical management of scarce resources (rooted the Mayflower Problem), the ethical obligations current generations have to future ones (inter- vs. intra- generational justice), and the ethical tensions embedded in policy decisions that favor resource conservation over investment in innovation, and vice versa (weak vs. strong sustainability). Each game will be underwritten by a game-theoretic structure that creates tension between individuals, who wish to achieve personal wellbeing through displays of egoism, and groups, who can only achieve communal wellbeing when cooperative decisions occur that promote collective good. By framing sustainability ethics problems in terms of non-cooperative game theory, in which the outcomes that impact each player are determined, in part, by the actions of other players, we aim to position students in explicitly social settings that require coordination of decision processes to ensure group success.

A vivid depiction of the link between game theory and sustainability that we are envisioning is captured in a political cartoon about the Prisoner's Dilemma that has the following narrative. The opening panel presents two well-dressed men in an office, with the first saying, "The crash is psychological. If everyone makes a leap of faith and starts spending again, we'll be fine!" The second replies, "What if no else starts shopping? You'll have sunk deeper into debt as your income shrinks." Determined to prove the skeptic wrong, the first man sets out to lead by example, but not before making a green analogy: "It's like recycling. If everyone does it, the world is a clean place." Unfortunately, the last panel depicts the man as unshaven and homeless, living on a garbage-strewn street, declaring, "Sure is a lot of litter out here." What the cartoon illustrates, then, is that whenever people make decisions that affect others, or act in response to actions, or even expected actions, of others, they are playing a game. In the case at issue, the homeless man made the mistake of selecting the cooperative strategy when the majority of other players choose to defect. Ultimately, what game theory studies and game-theoretic games capture are the outcomes that arise when rational individuals engage in strategic, interactive decisions. Ideally, playing iterative game-theoretic games will help students deal more effectively with game theoretic scenarios that arise in the real world, while also endowing them with a strong desire to prevent dangerous games from arising, and a solid base of conceptual recourses and experience to draw from, to propose viable alternatives.

The Externalities Game

Since "An Experiential Pedagogy for Sustainability Ethics" is at an embryonic stage, none of the games are fully developed. Nevertheless, it is possible to give readers of this volume some sense of the experiences playing them will engender. Because the externalities game is further along than the others, we can present a succinct summary here that captures the essence of what was conveyed during various early stage demonstrations.

The concept of "environmental externalities" is central to environmental economics. It typically refers to the result of industrial systems managers passing on the social costs of pollution to communities who do not directly participate in managing those systems. When environmental externalities are not dealt with appropriately, through, for example, adjustments in the market (e.g., externalities tax) or through government regulation (e.g., emissions caps, mandatory use of pollution abatement technology, termination of activity, etc.) the real cost of pollution goes unpaid, and injustice often goes unrectified, manifesting in the form of an unfair distribution of benefits and harms, as well as in other ways, including the abuse of power that occurs when elites take advantage of vulnerable populations.

Before playing the externalities game, students will be assigned primers that go into greater detail than the thumbnail sketch provided here. Some of the requisite material will include discussion of Nobel Prize winner Ronald Coase's seminal 1960 paper, "The Problem of Social Cost." In that essay, Coase defends a solution to the problem of externalities that subsequently became codified as the Coase Theorem: Under conditions where property rights are secure and transaction costs are low, the optimal course of action consists of the polluting and damaged parties negotiating a transfer of payments that either compensate the damaged parties (if pollution is continued the most profitable comparative outcome) or compensate the relevant parties associated with the offending industrial system for reducing their pollution (if that reduction is the most profitable comparative outcome).

After discussing the preparatory literature, students will be given the background and rules of the externalities game. The rules determine permissible moves in the game, and the background is the basic scenario that the game simulates. Because the scenario is a narrative description, obtaining it provides students with a concrete cognitive anchor. As an intelligible story, they can turn to it to make sense of the mathematically constrained options that structure game play.

In early testing, the following instruction sheet was used. It adapts a scenario from Coase's aforementioned paper.[2]

Rules:

- Players are randomly assigned one of three production roles: luxury, intermediate, and subsistence.
- Students must make production decisions that result in points (for themselves) and externality costs (shard by everyone). Whereas points accumulate linearly with production, costs expand exponentially.
- Luxury players gain the most points per unit of production, but also emit

---

2   Testing revealed that while this scenario seemed plausible during the initial development, in practice it does not work as well as envisioned. It would take us beyond the scope of this chapter to discuss the problems that we ultimately recognized. For present purposes, it is sufficient to note that both designing and playing the games involves experimentation. Unlike ethics projects that occur within the liberal arts horizon, this one does not move at the speed of philosophical imagination.

the greatest amount of externalities. Intermediate players gain the second most points per unit of production, and emit the second highest amount of externalities. Subsistence players gain the least amount of points per unit of production, but emit the least amount of externalities.

- Grades are determined by the points that are produced individually, minus the social costs that are produced collectively.
- A round of game play consists of two steps: 1) simultaneous production decisions by all players, followed up by, 2) simultaneous point-sharing decisions.
- Players can only produce whole goods up to their maximum and no less than 0.
- Players can share grade points after each round.
- Players can make deals during the game to share points or limit production for the greater good.
- Players do not have to share with anyone if they do not choose to. The people they share with and they amount they share is solely of their own discretion.
- Players can get above 100% as their final grade for each round, but they can get no less than a 0.

Background:

- Luxury Players: You are representing a railroad that carries coal to the city. You currently have enough coal to produce a maximum of 10 trips to the city and back. You know that if you make more trips, you can greatly increase your profits. The local farmers claim that your passing increases the damage done to their crops by fire and pollutants, decreasing their profits. At the same time, the cattle raisers say that your smoke is causing sickness to their cattle and reducing their profits as well. You must make enough trips to make a profit though.
- Intermediate Players: You are the cattle raisers. You currently have enough land to maintain a maximum of 55 cows. You want to keep as many cows as you can, though you know that if you raise too many, you will have a harder time keeping track of them. The railroad claims that their manure reduces their profits by eroding the track, and the farmers complain that they lose profits because their crops get eaten and trampled. You must decide how many cows you will maintain.
- Subsistence Players: You are the farmers. You currently have enough land to produce a maximum of 160 crops. You know that if you grow too many crops, the soil starts to soften, and the railroad tracks become unstable. The crops will also remove all the nutritional value from the grass, causing the neighbors cattle to be thin and weak. As your crops decline in health, you must purchase more resilient seeds, and nutritional supplements for

the fields, reducing your profits. You are tired of the railroad and factory taking your losses, but you don't seem to be able to produce enough profits to make a living. Perhaps you can persuade the other roles to share their wealth for the loss in profits that you are taking.

Play:

- Players are allowed to experiment with various production choices. The consequences of their production choices may not be fully attained until after the first round has been played, and if the instructor chooses, the first round can be considered a "practice round". Players will continue to experiment and try to make deals with each other to attain their wanted grades.
- To experiment with possible choices, players can enter their production decision into the computer and view the consequences of that decision. If the players do not want to cooperate with the other players, then they do not have to, but it is in their best interest to do so.
- When all players have come to a final decision, or when the instructor states that deliberation must end, the players will submit their final production value. *THIS VALUE DOES NOT HAVE TO BE WHAT WAS AGREED UPON WITH THE OTHER PLAYERS*
- The grades are then calculated by the instructor's program, and they are returned to the students with a list of emissions generated by each player. Another round may be played if the instructor so chooses.
- When all rounds have been played, the grades are calculated based upon an average of all grades through all rounds. Their final grade for this part of the game will be recorded and carried on through the next game.

To help students appreciate the operative mathematical constraints governing the externalities game, they also will be shown graphs, before game play commences, that illustrate three crucial points at a glance. 1) If the luxury players produce at maximum capacity and everyone else produces nothing, they can do very well; but if everyone produces at a maximum, they will fail. 2) If the intermediate players produce at a maximum capacity and everyone else produces nothing, they do very well; but, if everyone produces at a maximum, they will fail. 3) If the subsistence players produce at a maximum and everyone else does nothing, they can do well; but, if everyone produces at a maximum, they will fail. This information is displayed in a manner that concretizes the relation between production, externalities, and points for each producing class, and, crucially, emphasizes the risk of collective failure that will occur should any type of producer attempt to win through unilateral production.

In the testing we have done so far, students have enthusiastically experimented with different strategies in an attempt to discern whether egoistic behavior can correlate with success. In the midst of failures and conflicts that are experienced with emotional intensity, the following types of moral questions typically arise as constitutive elements of the deliberative discourse:

- What is an appropriate standard for determining how production should be allocated among the three different types of producers?
- Can moral arguments persuade people to act? That is, can moral reasons be the causes of desired actions? If so, can they be so persuasive as to be sufficient for preventing over-production by all players?
- Are any of production classes more responsible than others for taking leadership roles in proposing cooperative strategies?
- What standard should be used to determine trustworthiness?
- How can agreements between players be enforced?
- What consequences, if any, should arise if players do not uphold their promises?

Students have also demonstrated an aptitude for being reflexive, asking questions like the following queries into what the game potentially represents and the disanalogies between it and real world correlates.

- What do the different categories of production classes correspond to in the real world?
- What types of real policy decisions would best exemplify winning gaming strategies?
- Is it morally or pragmatically sufficient to address the problem of CO2 emissions as a matter of managing externalities and creating schemes for the just allocation of funds (potentially equivalent to points)?

Susan Spierre's Application

The concern over how to approach CO_2 allocation, as a form of climate change mitigation, is an example of an intrinsically moral problem that might be informed by experience with the Externalities Game. The complexity results from the fact that there are relatively few individuals benefiting from CO_2 emissions in the form of consumption, mainly by populations of wealthy, industrialized countries, and a growing population in low-income, less-developed countries, where most of the suffering associated with climate change is forecasted to occur. These low-income regions, who have contributed minimally to the problem, are already suffering from rising temperatures (e.g. from sea-level rise and/or agricultural setbacks), and typically lack the ability and resources to adapt. Consequently, there is an inequitable distribution of benefits as well as damages associated with climate change, and these clear imbalances illustrate the critical need for an equitable way to reduce CO_2 emissions globally. As the Externalities Game is calibrated, an analogous moral tension exists between the luxury and subsistence players, who must come to some sort of cooperative agreement about how grade points should fairly be allocated.

Moral philosophy literature provides an outline of carbon allocation methods that advocate the use of ethical and moral reasoning behind international climate policy, and typically advocate for an equal per capita distribution of the right to emit CO_2. Current policies, such as the Kyoto Protocol, promote a strategy of charging wealthier, developed countries with the responsibility to reduce emissions and the obligation to pay for the implications of climate change. While this approach is intuitively appealing from the standpoint of internalizing the external costs of greenhouse gas emissions, it fails to address the legitimate development needs of poorer nations. Even in cases of emissions trading, in which poorer nations are compensated for the emissions of wealthier nations, it remains unclear how transfer payments would improve the human condition. However, a more equitable method of allocation might be achieved by incorporating the *capabilities theory*, an approach where the focus of justice is not simply on distribution, but more specifically on how various goods are transformed into the capacity to flourish. The assumption that the welfare of a human being depends on one's consumption, captured in the economic neo-classical rational actor model, forces a trade-off between material consumption and environmental protection. Conversely, the capabilities approach suggests that such a compromise is an oversimplification by assigning value to the role of amenities in human well-being such as health services, educational institutions, and employment. Using a capabilities approach to measure human welfare, which is the goal of the United Nation's Human Development Index (HDI), is significant for climate mitigation policy because it removes the concentration on financial wealth, and takes into account other resources that are vital for people to live quality lives. The HDI, for example, includes three parameters in assessing human well-being: life expectancy, education, and income – either by establishing a broader understanding of human well-being (compared with consumption measures alone) that could establish an alternative moral basis for allocation of CO_2 emissions.

The distribution of CO_2 emissions for each country, established using the capabilities theory, is determined by solving an optimization problem of two functions: a human welfare function and a climate change damage function. The human welfare function is determined by a diminishing returns relationship between a countries' HDI value and its associated CO_2 emissions per capita. The global climate change damage function represents an exponential increase of potential damages to ecosystem services associated with increases in CO_2 concentration in the atmosphere. It is critical to assume a linear relationship between CO_2 emissions and atmospheric CO_2 concentration in order to relate both functions. The goal is to determine the point of greatest distance between the welfare and damage functions, which represents the optimal allocation of CO_2 emissions and related HDI value, considering the potential damages of climate change. Under this allocation method, countries emitting more CO_2 per capita than the optimal amount will be responsible to decrease emissions. Countries emitting less than the optimal level will have the right to increase emissions, but only with the intent to use those emissions to improve the capability of its citizens to live high quality lives.

In the context of CO_2 emission reductions, the implication of using a capabilities approach is that developed countries that emit more CO_2, may be able to reduce their consumption without necessarily reducing their overall quality of life. Also, less developed nations that emit less, have an incentive to focus development efforts on improving the lives of its citizens, rather than increasing consumption of material goods. Incorporating the capabilities approach in allocating CO_2 emissions is a more equitable method than previously contemplated methodologies because if successful, this technique would result in a widespread improvement in human well-being. After experiencing the Externalities Game for themselves, we theorize that students will be able to reason through moral questions such as climate change with greater facility—which is not to suggest that a capabilities approach is necessarily the "correct" resolution of the game. Rather, we suggest that engineering & science students will be better prepared to engage in the literature and deliberative processes of climate policy, formulate (and test, to some extent) alternative moral hypotheses, and make stronger connections between the simplified moral questions they have confronted for themselves in class, and real situations they will be confronted with as professionals.

References for Further Discussion

Thomas P. Seager and Evan Selinger, "Experiential Teaching Strategies for EthicalReasoning Skills Relevant to Sustainability." Proceedings of the 2009 IEEEInternational Symposium on Sustainable Systems and Technology: http://ieeexplore.ieeeorg/xpl/freeabs_all.jsp?tp=&arnumber=5156721&isnumber=5156678.

Thomas P. Seager, Evan Selinger, Daniel Whiddon, David Schwartz, and Andrew Berady,"Debunking the Fallacy of the Individual Decision-maker: An Experiential Pedagogy for Sustainability Ethics." Proceedings of the 2010 IEEE International Symposium onSustainable Systems and Technology. Online: http://ieeexplore.ieee.org/xpl/freeabs_all.jsp?arnumber=5507679

Susan Spierre, Thomas P. Seager, and Evan Selinger, "Determining an Equitable Allocation of Global Carbon Dioxide Emissions." Proceedings of the 2010 IEEE International Symposium on Sustainable Systems and Technology. Online: http://ieeexplore.ieee.org/xpl/freeabs_all.jsp?arnumber=5507704

Section 3
Culture

Introduction to Part 3:
Climate Change and Culture

Søren Riis

Climate change will lead to extraordinary cultural and social changes which will also bring the human and social sciences to a breaking point. In this section, the researchers self-reflectively struggle with the challenges posed by climate change. Climate change cannot be sustainably mitigated or adapted to without the help of the social and human sciences, yet there are no easy fixes available. Instead, this section raises a number of important concerns.

In their chapter, "Nature, Climate Change and the Culture of Social Sciences", Reiner Grundmann, Markus Rhomberg and Nico Stehr focus on the historic invisibility of social scientists in debates concerning climate change. The authors address the fundamental challenge and ambivalence of social scientists in having to learn to address climate change. The social sciences face major challenges in coping with climate change as they strive to avoid ecological determinism, and at the same time support their disciplinary founding within social constructivism. In addition, they must be aware that their criticism against ecological determinism may immediately be politicized and misused by fossil fuel lobbyists.

Andrew Jamison also identifies a shortcoming in academia in the general failure to address and understand the relation between the emergence of climate change knowledge and the rise of social movement theory. In his chapter, "Climate Change Knowledge and Social Movement Theory", Jamison unfolds a definition of a social movement and elaborates on the origin of the environmental movement of the 1970's. From here he shows how the climate change movement has been influenced by neo-conservatives, who tried to obstruct the movement and how neo-liberals instead were capable of connecting climate change to green business and thus turned climate change into a discourse concerning technological innovation. Based on these historical accounts and analyses, Jamison shows that climate change knowledge-claims emerging out of the heterogeneous climate change movements call for a nuanced epistemology. Jamison's paper gives us a clear idea of what such an epistemology must be able to account for.

The unrest in Gert Goeminne's chapter, "The matter of climate change: What it is and how to be concerned with it" stems from what he sees as a crucial shortcoming in current environmental policy. There is a clear need to develop a conceptual space which allows policymakers and scholars to better address nature-culture hybrids. Without such a conceptual framework, new environmental policies are destined to fail. Inspired by Bruno Latour, Goeminne argues that the environment must be explicitly framed as a culturally reflected "matter of concern".

Joshua Forstenzer shows the plurality of problems arising from climate change, paying special regard to the cultural and political shifts caused by it in his chapter, "Education, Active Citizenship and Applied Social Intelligence: some democratic tools to meet the threat of climate change". Forstenzer's perspective of climate change reinvigorates John Dewey's theory of democracy. Instead of calling on experts to find a sustainable solution to climate change, Forstenzer together with Dewey, argues that an adequate cultural change in a liberal democracy may only be brought about by means of an active sense of citizenry. However, the paper ends on a sceptical note, as it raises serious doubt as to whether broad social and cultural educational measures may indeed be timely enough to address the acute problems at stake.

In Michel Puech's chapter, "Sustainability Means Ethics and This is a Cultural Revolution", a doubt and lack of trust in broad institutional approaches to climate change is deepened. In particular, the years of preparation and ultimate failure of the COP 15 has demonstrated the inadequacy of the UN in dealing with climate change. Puech takes this failure as an *a posteriori* argument for having to bypass institutions in the quest for finding sustainable solutions to climate change. Puech argues in favour of what he calls an ethical revolution that is radically preoccupied with micro-politics and individual lifestyles. This revolution has to be deep enough for 'an ethical self' to emerge – only this way one can hope for making any real progress in mitigating climate change.

The last chapter of the section, "Coping with Climate Change: Social Science and the Case of Multi-site Living, by Jørgen Ole Bærenholdt, partially takes the shape of a thought experiment. After addressing how different social scientists have recently responded to climate change, Bærenholdt focuses on change in practice and takes the case of multi-sited living seriously. Based on multi-sited living he argues that we may be able to discover the contours of a new and untraditional way of dealing with climate change – a way that connects mobility and sustainability studies.

Chapter 3.1

Nature, Climate Change and the Culture of Social Sciences

Reiner Grundmann, Markus Rhomberg and Nico Stehr

Introduction[1]

At the 2008 Annual Conference of the British Sociological Association, John Urry gave a plenary talk on climate change in which he made the strong case that sociologists so far had not engaged very much with the topic of climate change and that it was high time to get involved. In his concluding words he spoke of a "call to arms" for Sociology. The urgency and the political nature of climate change should lead us to think about our faults, omissions, and options carefully. However, sociologists should not rush into the discursive arena without asking some critical questions in advance, questions such as: what exactly could Sociology contribute to the debate? And, is there something we urgently need that is not addressed by other disciplines or by political proposals?

In this chapter we examine the question of why social scientists have been fairly invisible in professional circles as well as in public and policy discussions about global warming. Our argument focuses on two reasons: The first is sociology's legacy of social constructivism (which we would like to defend) and its skepticism of ecological determinism, including climate determinism. The second is the politicization of the climate change debate and a problematic reaction from sociological community.

Hence the chapter is organized as follows:

(1) As anthropogenic climate change has evolved from a science-based issue to a top global policy issue, social scientists have developed a mixed but largely uneasy relationship with the topic. This is due to the politicization of the debate. Science studies scholars, for example, who were in principle well-placed to contribute to the debates, felt uneasy in a polarized debate where academic research might be seen as politically counterproductive. (2) Following the argument above, we refer to the notion of framing as one of the key sociological insights when dealing with policy-relevant information. Framing rejects the assumption that there is anything self-evident to be found "out there". The framing of climate change as

---

1   A closely related publication was published by Reiner Grundmann and Nico Stehr (2010): Climate Change: What Role for Sociology? A Response to Constance Lever-Tracy. *Current Sociology*, 58(6), 897–910.

a social and political issue is therefore likely to be among the prime concerns for sociologists researching this area. (3) We conclude with suggestions why and how social science discourse should overcome its reluctance to move into the center of scientific and policy debates about the ways in which societies have to live with climate risks.

Thematic Frame

This chapter seeks to develop some considerations on our questions: does the dramatization of events lead to effective political responses? Do we need a politics of fear? Is scientific consensus instrumental for sound policies? And more generally: what are the relations between a changing technological infrastructure, social shifts, and belief systems? How was it possible that the *fight against climate change* rose from a marginal discourse to a hegemonic one (from Heresy to Dogma)? And will the discourse remain hegemonic or will long and intensive public debates about climate change lead to *climate change fatigue*?

The urgency of the matter is becoming more apparent as evidence mounts that efforts to date to mitigate greenhouse emissions are falling short of what is required to *arrest* or *slow* climate change (see Matthews & Caldeira 2008), that global emissions are likely to rise significantly in the next years (Sheehan 2008), and that transnational policies and treaties put in place to reduce emissions fail to reach their targets, (let alone achieving the objectives needed to put a brake on climate change; see Prins & Rayner 2007).

The inherent alarmism in many social science contributions on climate change merely repeats the central message provided by mainstream media. It is curious that little guidance is provided in terms of what could be done to deal with climate change, or at least pointing out what options we have. Can Sociology deliver on this? One would expect substantial proposals besides the exhortation to listen more to what the natural sciences have to tell us. As Phil Macnaghten and John Urry (1998: 15) put it:

> "Once we acknowledge that ideas of nature have been, and currently are, fundamentally intertwined with dominant ideas of society, we need to address what ideas of society and of ordering become reproduced, legitimated, excluded, validated, and so on, through appeals to nature or the natural. And the project of determining what is a natural impact becomes as much a social and cultural project as it is 'purely' scientific". Interactions with the global environment are ultimately grounded in a wealth of underlying social factors and historical contexts".

If the political urgency of the matter is one thing, Sociology's contribution is another. Sociologists have been fairly invisible in professional circles as well as in public and policy discussions about global warming.

But why? We argue that science studies scholars, for example, who were in principle well-placed to contribute to the debates, felt uneasy in a polarized debate where academic research might be seen as politically counterproductive.[2] This leads us to the crucial question of Sociology's core contribution to the problem. It has been said many times that Sociology as an academic discipline came into being at a time when Western societies were making a transition from agricultural societies to industrial societies. This change led to enormous shifts in social structure and cultural values. Both of these core elements were seen as influencing each other and are still today Sociology's core program: the study of social change together with technological change, change in social relations and cultural systems. We are now witnessing global change that is at least of a similar order of magnitude as the earlier change. Some societies in developing countries make the transition to a less agricultural, more industrialized type while a part of its workforce is migrating around the globe. The workforce in developed countries is encouraged, even forced to participate in higher education and to adopt a lifestyle that is flexible enough to manage transitions in the labour market. From this follows that we have a more educated population, sometimes eager and ready to participate in complex decision making, while at the same time trying to cope with cultural value systems they have inherited from an earlier epoch. Sociology has been one of the leading disciplines in the study of globalization. The question is if it can do the same in the study of climate change.

Posing the question this way already suggests an answer. While Sociology was extremely well placed to detect the process of globalization as a social change issue, climate change was from the beginning the domain of natural scientists who addressed and framed the issue according to the parameters of their specific disciplines. Dominated by the modeling community, the study of climate change largely, though not exclusively, amounted (and still amounts) to the development and refinement of global circulation models that depend crucially (among other things) on sound input data. Climate change modelers are keen to get usable data from other academic communities but only if these are in the "right format", so to speak. They want cost of life estimates from economists and predictions about mitigation and adaptation costs. They would like from sociologists some guidance about the relation between climate induced stress on social communities and the likelihood of wars or revolutions. Such estimates would be valuable for predictions of further greenhouse gas emissions and their abatement. Few sociologists would subscribe to the notion that social processes can be 'modeled' in this way (or even

---

2   Brian Wynne (1996: 363, 372) put it this way: "A sociological deconstruction of knowledge may find itself in unwelcome company, politically speaking. (…) These sociological observations about the scientific knowledge of global warming could of course contribute to a deconstruction of the intellectual case for the environmental threat, and thus also to a political demolition of the 'environmentalist' case for internationally effective greenhouse gas controls."

that it should be a prime task of sociologists to do so).[3] Since few sociologists are doing so, let alone engage in interdisciplinary co-operation with climate scientists, the latter are literally encouraged to become social scientists themselves. The outcome of such cognitive shifts is very likely technocratic perspectives and policy advice. Scientists and engineers have acted as 'lay sociologists' on many an occasion and have – based on their personal beliefs and values – built several assumptions about human behavior into their theories.

The main issue that remains therefore is the question: why has mainstream sociology failed to incorporate global warming within its horizon? As we pointed out, we see two reasons for this. The first is Sociology's legacy of social constructivism and its skepticism of ecological determinism, including climatic determinism. The second is the politicization of the climate change debate and a problematic reaction from the sociological community. We address both in turn.

Politicization and Framing of the Climate Change Debate

The grand narrative of climate refers to stories (found in scientific literature and in public discourse) in which descriptive and prescriptive elements are liberally mixed. They inform us about the rise of civilizations, the fate of nations and the character of humans under favorable or adverse climatic conditions. "Permissive" or "harsh" environments of a geographical area determine not only the material but the immaterial social and cultural realities. And, as the French historian Edmond Demolins, at the turn of the last century, confidently asserts, "if history of mankind were to begin over, without any change in the world's surface, it would broadly repeat itself" (as cited by Landes 1998: 533). Unfortunately for the grand narrative of climate, different cultures emerge in similar natural environments and similar cultures in different natural environments. What is more, anthropogenic climate change alters the climatic conditions in historical times and global warming may change climates within the time span of a generation.

Politicians often claim that their climate policies are made on the basis of scientific information. It is a common perception that the more knowledge we have, the better the political response will be. This lies at the heart of the linear model of policy making that has been dominant in the past albeit debunked time and again (Godin 2006). What we increasingly realize is that knowledge creation

---

3   The same cannot be said for members of other social science disciplines, for example for economists, anthropologists or geographers. The significant role of economists in discussions of the consequences of climate change and climate policies (e.g. Stern 2006) may also be one of the reasons why sociologists are at best marginal voices in climate change discussions. Economists have dominated debates on the appropriate responses to the climate change by successfully defining such responses as economic issues, for example, how costly will it be to mitigate greenhouse gas emissions and what economic measures may be put in place to encourage energy efficiency (carbon taxes, trading).

leads to an excess of information and "objectivity" (Sarewitz 2000). Even the consensual mechanisms of the Intergovernmental Panel on Climate Change lead to an increase in options because knowledge about climate change increases.

Scientific findings are often used despite all ostensive appraisal, but for a particular reason, namely to justify positions. This implicates that the political system is not dependent on scientific findings. It only uses them to justify decisions it had already taken. You can always find an expertise and a contra-expertise that suits your interests. In a network analysis on transnational policies on the ozone hole, Grundmann (1999) refers to the political instrumentalization of scientific findings. The political system will not give up its power of decision-making, at least it will demonstrate this constantly in the public. Actually, one can observe the tendency, that political communications influence the scientific debate. Weingart (2001, 2003) suggests a *Politicization* of scientific research: By discussing different scientific expertise in the political system, these debates are extended into the scientific sphere and politicize the scientific debate too.

Ulrich Beck (1992, 2007) describes in his concept of risk society a mediatization and politicization of the scientific system as part of the modernization process: Until the first half of the 20th century scientists were engaged in inventions about nature, human beings and society. But since the 1960s and 1970s they are forced into a new frame: in a reflexive phase of research, science has to deal with its prior findings and its consequences. Since then it is confronted with its own products and defects. This has also influenced the relationship between science and its social environment. While in the former phase the science was celebrated by the public as a successful actor, in the latter it is confronted by political actors, the media and the public with its faults from the past, which became obvious as follow-up problems. As a consequence, the science lost its monopoly as key definer of risks. This is ambivalent: On the one side, a modern and complex society requires more research for its social progress, but on the other hand, because of the public perception of scientific uncertainty, science has lost its unique position to define social problems and risks in modern societies. Other social actors, formerly the receivers of scientific findings, became active definers. So, new definitions are not primarily based on science alone but on broader social concerns. Therefore, the quality of risks is based on its social construction. Especially risks, which are withdrawn from our direct cognition, like radioactivity, harmful substances and obviously the potential impacts of climate change, are based on social constructions. By following this logic, that our understanding of climate and climate change is a social construct, we also have to consider theories lying under this assumption.

It is an often repeated myth that constructivism ultimately means playing into the hands of climate change deniers. One of the key sociological insights when dealing with policy-relevant information is the notion of framing. This is a social constructivist concept that rejects the assumption that there is anything self-evident to be found "out there". The framing of climate change as a social and political issue is therefore likely to be among the prime concerns for sociologists researching this area. Frames are used in the sense of organizing devices that allow the selection and emphasis of topics to decide "what matters" (Gitlin 1980). Gamson and Modigliani (1989) consider frames as being embedded within "media packages" that can be seen as "giving meaning" to an issue. According to Entman (1993), framing comprises a combination of different activities such as: problem definition, causal interpretation, moral evaluation, and/or treatment recommendation for the item described (see Grundmann and Krishnamurthy 2010)

Taking a closer look on the Kyoto protocol, Prins and Rayner (2007: 8) notice that the Protocol is "a manifestation of a particular framing of the climate change issue". Kyoto is already defined as news, since it stands for years as the particular global action plan for climate policy. The Kyoto protocol almost exclusively deals with questions of mitigation as reduction of greenhouse gases. So, one can observe clear thematic selection processes. Post (2008) has shown, that German news media present foremost stories about the reduction of greenhouse gases, as for example codifications for emission-trades, emission-taxes, or emission-reducing technologies, but they hardly report stories with the topic of adaptation on climate change (see also Boykoff & Timmons 2007). Kyoto is linked to mitigation in terms of the media concept of news values (Galtung & Ruge 1965), especially one can observe the value of continuity: People know the Kyoto protocol and have a vague idea about the topics, mitigation and reduction, which were discussed there. Hence mass media can refer on this prior knowledge (Fredin et al. 1996, Price & Zaller 1993).

Framing describes the selection, exclusion and accentuation of specific attributes by which the content of a topic is contextualized, or framed, by the news media. By choosing a certain perspective in the news coverage, the evaluation by the audience of the topic is navigated consciously or unconsciously in a specific direction. But the audience is not entirely dependent on the media exposure: Individual experiences as well as social knowledge launch these interpretations (Entman 1993). As Trumbo and Shanahan (2000: 201) argue, cycles in media coverage embody narratives that guide public understanding of climate change: "The relationship between the content of the mediated information environment and the state of public understanding can be especially strong for (…) issues that have low intrusiveness in the present, have distant time horizons, and have high levels of conflict. Climate change embodies all of these characteristics".

Sociologists have indeed studied the emergence (or construction) of scientific claims as facts (see, for example, Demeritt et al. 1998, Lahsen 2005, Skodvin 2000). To define global climate change as a social construction is not to diminish its importance, relevance, or reality. It simply means that sociologists study the process whereby something (like anthropogenic climate change) is transformed from a conjecture into an accepted fact.

Many social scientists and scientific experts more generally are aware of the dangers that arise from a politicized environment. If it is true that controversies abound in risk debates, this leads to a dilemma for researchers in the field. On the one hand, they may want to see the problem solved according to their value preference, no matter how strong the scientific evidence is. On the other hand, they may be very reluctant to be drawn into a controversy, which could harm the integrity of scientific research. Many scientists believe that it is imperative to stop global warming and therefore critical voices within the academic world (or in public discourse) are seen as a hindrance. Because the political project of GHG mitigation represented by the UNFCCC, supported by the IPCC, and resulting in the Kyoto Protocol, is for many the only game in town, other options have been pushed off the agenda. The perception was, and still is, that critical discussions of the IPCC and the Kyoto framework would lend support to the contrarian forces who wanted to derail any political project addressing climate change. It may be that this constellation had an impact on Science Studies, and there has been very little research on this topic in recent years. A search for the key words "climate" or "warming" in the titles of articles appearing in Science, Technology and Human Values and Social Studies of Science yields just 14 articles from 1992 to 2008. No doubt, many scholars working in this field are sympathetic to environmental values and thus uncomfortable being seen as part of the sceptical camp. But there is a deep irony here. One of the methodological principles of Science Studies is to study controversy from a standpoint of symmetry, i.e. not privileging one account over the other. The so-called Strong Programme (Bloor 1976) holds that science studies scholars should remain neutral with respect to the truth claims science makes. They should explain the success or failure of a scientific theory in the same terms. The outcome of all scientific controversies – successful or not – should be explained by social factors. As the climate change controversy unfolded in the United States, journalists seemed to follow a similar principle in their reporting, thus giving a boost to the small number of climate change sceptics (Boykoff & Boykoff 2004). Could it not be that this political climate prevented sociologists from analyzing the science and politics of climate change to any great extent? Might it not be that the political and social forces have become so strong that the very principles of the field of science studies have taken a back seat? In other words, we argue that there has been little engagement with climate change on the part of sociologists (and especially science studies scholars) because they are aware of the political implications and anxious of not wanting to play into the hands of climate change deniers.

Conclusions

It may become more frequent that core Sociology journals will publish articles related to climate change. But what would editors be looking for? Probably for research results that are based and interpreted along the lines of the sociological canon or in critical discussion with it. In institutional terms this would mean to give more prominence to subfields like environmental sociology, SSK and STS studies and to open the theoretical debate about the core conceptual framework and identity of the discipline (see, for example, the sympathetic reception of Bruno

Latour as a keynote speaker at the 2007 BSA conference). In a way, some of the 'marginal fields' could move towards the centre of the discipline. No doubt, there will be limits to such a move that ironically has to do with a certain reluctance of these fields to carry out their own programme. By this we mean the concern that anything that could be seen to cast doubt on the 'integrity' of the climate scientists has to be avoided in order to protect the political impetus behind it. There seems to be the curious conviction that lest you want to be accused of helping the fossil fuel lobbies and the climate sceptics, you better keep quiet.[4] In addition, these marginal fields would need to embrace the idea to move closer to the centre of sociology, an aim which cannot be taken for granted. Anthropogenic climate change is an example of a dialectical relationship between nature and society and therefore ought to move into the center of sociological concern. However, there lingers the question of what is an essential sociological methodology.

Efforts to protect the climate from society (mitigation) and measures to protect society from the climate (adaptation) both are efforts that have at their core the assumption that social conduct can alter natural processes and that natural processes can have a significant impact on social relations. From this it follows that sociologists should study the direct interchange between society and natural environments as mediated through technology (sociology of science and technology), the relation between knowledge and decision making, the values that form our dealing with nature (sociology of culture), the social shifts that contribute or mitigate climate change, and the consequences a changed climate will have on social, economic and political affairs. Most importantly, if Sociology wants to provide practical knowledge, then it needs to identify the social structures and dimensions of agency that will facilitate transitional pathways to a low carbon society.

References

Beck, U. 1992. Risk Society. Towards a New Modernity. Cambridge: Polity.

Bloor, D. (1976). Knowledge and Social Imagery. 2nd edition. London: Routledge.

Boykoff, M., and Boykoff, J. 2004. Balance as Bias: Global Warming and the US Prestige Press. Global Environmental Change, 14(2), 125–36.

Boykoff, M., and Timmons, R.J. 2007. Media Coverage of Climate Change: Current Trends, Strengths, Weaknesses. Human Development Report 2007/2008: Human Development Report Office.

Demeritt, D. 2001. The Construction of Global Warming and the Politics of Science. Annals of the Association of American Geographers, 91(2), 307–37.

---

4   Confronted with conspiracy theories after 9/11, Latour (2004) has become wary of too critical an attitude towards climate science: "Why does it burn my tongue to say that global warming is a fact whether you like it or not? Why can't I simply say that the argument is closed for good?"

Entman, R. M. (1993). Framing: Toward Clarification of a Fractured Paradigm. Journal of Communication, 43(4), 51–8.

Fredin, E. S., Kosicki, G. M. and Becker, L. B. 1996. Cognitive Strategies for Media Use During a Presidential Campaign. Political Communication, 13(1), 23–42.

Galtung, J. and Ruge, M. H. 1965. The Structure of Foreign News. Journal of International Peace Research, 2, 64–90.

Gamson, W. A. and Modigliani, A. (1989). Media Discourse and Public Opinion on Nuclear Power: A Constructionist Approach. American Journal of Sociology, 95, 1–37.

Gitlin, T. 1980. The Whole World Is Watching: Mass Media in the Making and Unmaking of the New Left. Berkeley, CA, Los Angeles, CA & London, U.K.: University of California Press.

Godin, B. 2006. The linear model of innovation. Science, Technology & Human Values, 31(6): 639–667.

Grundmann, R. and Krishnamurthy, R. 2010. The Discourse of Climate Change: A corpus-based approach. Critical Approaches to Discourse Analysis across Disciplines, 4(2): 113–133.

Grundmann, R. 1999. Transnationale Umweltpolitik zum Schutz der Ozonschicht. USA und Deutschland im Vergleich. Frankfurt am Main, New York: Campus-Verlag.

Lahsen, M., 2005. Seductive Simulations? Uncertainty Distribution Around Climate Models. Social Studies of Science, 35(6), 895–922.

Landes, D. S. (1999). The Wealth and Poverty of Nations. Why some are so rich and some so poor. 2nd edition. New York: W.W. Norton.

Latour, B. 2004. Why Has Critique Run out of Steam? Critical Inquiry, 30, 238–239.

Matthews, H. D. and Caldeira, K. 2008. Stabilizing climate requires near-zero emissions. Geophysical Research Letters, 35(L04705), doi:10.1029/2007GL032388.

Mcnaghten, P. and Urry, J. 1998. Contested Natures. London: SAGE.

Post, S. 2008. Klimakatastrophe Oder Katastrophenklima? Die Berichterstattung Über Den Klimawandel Aus Sicht Der Klimaforscher. Frankfurt am Main: Verlag R. Fischer.

Price, V. and Zaller, J. R. 1993. Who Gets the News? Alternative Measures of News Receptions and Their Implications for Research. Public Opinion Quarterly, 57, 133–64.

Prins, G. and Rayner, S. 2007: The wrong trousers. Radically rethinking climate policy. James Martin Institute for Science and Civilisation: Oxford University.

Sarewitz, D. 2000. Science and Environmental Policy: An Excess of Objectivity, in Earth Matters: The Earth Sciences, Philosophy, and the Claims of Community, edited by Frodeman, R. Upper Saddle River, NJ: Prentice Hall, 79–98.

Sheehan, P. 2008. The new global growth path: Implications for climate change analysis and policy. Climatic Change, 91(3–4), 211–231.

Skodvin, T. 2000. Structure and agent in the scientific diplomacy of climate change: An empirical case study of science–policy interaction in the Intergovernmental Panel on Climate Change. London: Springer.

Stern, N., 2006. The Economics of Climate Change. The Stern Review. London: H.M. Treasury.

Trumbo, C. W. and Shanahan, J. 2000. Social Research on Climate Change: Where Have We Been, Where We Are, and Where We Might Go. Public Understandings of Science, 9, 199–204.

Weingart, P & Stehr, N. 2000. Practising Interdisciplinarity. Toronto: University of Toronto Press.

Weingart, P. 2003, Wissenschaftssoziologie. Bielefeld: transcript-Verlag.

Weingart, P. 2001. Die Stunde Der Wahrheit? Weilerswist: Velbrück Wissenschaft.

Wynne, B. 1996. SSK's Identity Parade: Signing-Up, Off-and-On. Social Studies of Science, 26, 357–91.

Chapter 3.2

Climate Change Knowledge and Social Movement Theory[1]

Andrew Jamison

Introduction

In all of the voluminous commentary that accompanied the failure to reach an international agreement on climate change in 2009, the connections between social movements and those who took part, either in the formal negotiation process or in the broader public debate, have tended to be neglected. And yet, both the negotiations and the public debate about climate change have been shaped, in significant ways, by social movements.

It was in the environmental movements of the 1970s that the idea that human activity could be changing the earth's climatic conditions first left the circumscribed confines of academic discussion to enter into the broader realms of society and politics. The environmental movements provided a social context, a cultural space for biochemists, ecologists and other natural scientists and engineers to educate the public about environmental issues. And, as part of that process of public education, the scientific conjecture that the accumulation of certain gases in the atmosphere, in particular carbon dioxide, could, as in a greenhouse, raise the temperature of the earth and literally warm the globe started to become public knowledge (Commoner 1971, Ward and Dubos 1972, The Ecologist 1972).

Today, climate change knowledge is a field of contention, with fundamental disagreements over the causes and the appropriate ways to deal with it (for a recent overview of the debate, see Malone 2009, and for a discussion of the reasons for the disagreements, see Hulme 2009). And while there are a great many different viewpoints, it can be suggested that there are three main positions in relation to climate change knowledge, which will be characterized here as dominant, residual, and emergent, borrowing a terminology developed by the literary historian cum cultural theorist Raymond Williams to understand processes of social change (Williams 1977). I have previously used these terms to explore the historical

1   This chapter was first published in Wiley Interdisciplinary Reviews: Climate Change. Climate Change Knowledge and Social Movement Theory, Volume 1, Issue 6, November/December 2010, Pages: 811–823. Copyright 2011 John Wiley & Sons, Ltd., A Wiley Company. This material is reproduced with permission of John Wiley & Sons, Inc.

tensions in the "making of green knowledge" between commercial, academic and critical orientations (Jamison 2001).

The dominant position is associated with those who have been most active in raising political awareness about climate change in the past decade, and who have promoted a substantial lowering of the emissions of carbon dioxide into the atmosphere and a transition to what has been termed a "low-carbon society". The residual position is associated with the self-proclaimed "skeptics", who, for various reasons question the importance of dealing with climate change as opposed to other issues, and have actively challenged the dominant position, primarily by questioning the truth value of the scientific knowledge claims that have been made in its behalf. The emergent position is associated with those who are convinced that climate change is occurring, and that it will have serious consequences if it is not abated, but who stress the importance of dealing with climate change in ways that take issues of justice and fairness seriously into account. The positions are neither mutually exclusive nor all-encompassing, but as ideal-typical categories, they can be helpful for exploring the connections between social movements and climate change knowledge.

It is the contention of this chapter that these three contending positions have been shaped, most especially in regard to their conceptions of science, by social movements. The neo-conservative, or neo-nationalist movements that emerged in the 1970s and grew to political significance in the 1980s, especially in the United States, but also in many European countries, have had a major influence on the development of the skeptical, or residual position, both in relation to climate change, in particular, and environmental issues in general (Helvarg 1988, Rowell 1996, McCright and Dunlap 2003, Jamison 2004). In these movements, adherence to conservative values or beliefs, including a traditional, discipline-based conception of scientific knowledge, has encouraged rejection of the findings of the climate scientists, who, together with their political spokespersons, such as Al Gore, have been of such central importance in making climate change an issue of public debate during the past two decades.

Many of the most vocal climate change debaters, particularly Al Gore himself, can, in turn, be considered to have close connections to the rise of neo-liberal, or "transnational capitalist" movements that grew into significant actors in the global political economy after the fall of the Soviet empire in 1989 (Sklair 1997). These movements have been especially important in promoting the establishment of closer relationships between academic scientists and business firms, and in commercializing scientific knowledge throughout the world, but most effectively in the United States, where the process of "academic capitalism" has been most pronounced (Bok 2003, Slaughter and Rhoades 2004). The conception of science that is shared by most of the vocal climate change debaters is "transdisciplinary" and entrepreneurial, in that the knowledge that is made in many climate research centers and panels is dependent on the contexts in which it is made, both financial and organizational (Gibbons et al 1994, Elzinga 1996, Hunt and Shackley 1999, Yearley 2009).

In recent years, it has been the so-called anti-globalization movement that has provided the social context for concerns with "climate justice" to be articulated as a third mode of climate change knowledge (Della Porta, ed 2007, Chawla 2009). Like the anti-globalization movement as a whole, however, the proponents of climate justice have yet to articulate a coherent sense of collective identity or common purpose, and that is why this position in relation to climate change knowledge is best characterized as emergent.

These relations between social movements and climate change knowledge have not received much attention either in the academic or more popular literature. Those who write about social movements tend to define and analyze them primarily in political and organizational terms, while those who write about climate change knowledge have tended to relegate social movements to the contextual background, focusing most of their attention on the specific discourses, deliberations and cognitive claims of actors and institutions concerned with climate change. As such, an understanding of the social movements that have served to help shape climate change knowledge has been neglected. While we all know intuitively that there are connections between broader political and social movements and climate change knowledge, the links are seldom focused on explicitly either in the scholarly or popular literature.

By reviewing the history of climate change knowledge from the perspective of social movements and social movement theory, and thus bringing, as it were, the contextual background into the textual foreground, this chapter attempts to help fill this gap in understanding.

What is a Social Movement?

There is little agreement among those who study social movements about what a social movement "really" is. Definitions depend on which movements are seen as typical or most important, what kind of terminology or conceptual framework is applied, the particular research questions being addressed and, not least, the situation, or standpoint of the researcher (Eyerman and Jamison 1991, della Porta and Diani 2006). Since Neil Smelser (1962) in his classic work, *Theory of Collective Behavior*, divided movements into "results-oriented" and "values-oriented" there has been a bifurcation among the students of social movements, namely between those who focus on what might be termed movements in the streets versus those who focus on movements in the mind. Even so, for all their differences, the various definitions that have been most actively applied in the scholarly literature can be said to share certain common features or elements which will be used to provide a "working definition" of social movements for the purposes of this chapter.

On the one hand, a social movement will be defined as a collective form of social behavior that is explicitly organized for political action. A social movement is the process by which human and material resources are mobilized to try to affect political change (an influential recent discussion is McAdam, Tarrow and Tilly 2001). They make use of what Herbert Kitschelt (1986), in his analysis of anti-nuclear movements in the 1970s, termed "political opportunity structures" in pursuit of one or another political cause. Social movements tend to manifest themselves through publicly recognized forms of protest, or direct action, but those acts do not themselves make a social movement. They need to be linked or connected to one another in some way, organized and coordinated by means of a common platform or program. This provides what has been termed a "collective identity", a set of values or precepts or beliefs that empower those who share them and identify with them. These matters of collective identity formation have been most influentially theorized and studied by Alain Touraine and Alberto Melucci and their students and colleagues (see, for example, Touraine 1981; Touraine 1988; Melucci 1985; Melucci 1996).

On the other hand, a social movement is something distinct from more formalized political parties, social institutions, or other established kinds of politics, in that it "moves", or to put it another way, it has a more informal character than established political and social activities; in this regard, social movements resemble what Ulrich Beck (1992) has termed "sub-politics". Many theorists have attempted to distinguish movements from institutions and movements from organizations and political parties (perhaps most famously Alberoni 1984), and a recurrent theme in social movement studies has been to analyze processes of institutionalization and professionalization. In recent years, social movements have come to be ever more likened to, and conceptualized as, networks and seen as central to what Manuel Castells has termed the "network society" (Castells 1996), and attention has been given to distinguishing between the different kinds of networking, brokerage and (inter)mediation that goes on in social movements (e.g. McAdam and Diani, eds 2003). In many contemporary social movements, not least those related to climate change, much of this networking activity is now conducted "virtually" via electronic communication and the Internet (Garrett 2006).

A third element of almost all theories of social movements is the understanding that social movements are more than "merely" political phenomena, and that they involve some form of what Jürgen Habermas (1984, 1987) has so influentially termed communicative action. As in other fields of social science, there has been an increasing interest in these more communicative, or cultural, aspects of social movements, even though there is little agreement about how best to study or theorize about them (cf. Buechler and Cylke, eds, 1997; Goodwin and Jasper, eds 2004). Many social movement analysts use the concept of framing to discuss these matters (Snow and Benford 1988; Benford and Snow 2000) while for others, the role of passions and emotions has come to be emphasized (cf. Goodwin et al, eds 2001). Other theorists, usually within anthropology, have focused explicit attention on the uses of knowledge by social movements, perhaps most influentially the uses of so-called indigenous knowledge by social movements in non-Western countries (e.g. Aparacio and Blaser 2008; Escobar 2008)

The cognitive approach to social movements (Eyerman and Jamison 1991) makes use of the terms "cognitive praxis" and "movement intellectuals" to emphasize the role of knowledge-making in social movements, and to characterize the people who are most actively involved. Cognitive praxis is defined as the linking, or integration of ideas, ideologies, and/or world view assumptions (a cosmological dimension) to particular activities or forms of action, including technical development, information dissemination and practical demonstration of both protest and constructive alternative (a technological dimension). The movement is seen as providing an organizational dimension, a public space, for integrating the cosmology and the technology in processes of collective learning, and it is their cognitive praxis that makes social movements particularly important in the constitution and reconstitution of science and technology (Jamison 1988, Jamison 2006). In a Eyerman and Jamison (1998) broaden the cognitive approach to encompass cultural practices and, by so doing, focus attention on the role that the "mobilization of tradition" plays in the collective activities of many social movements. It is the mobilization and (re)invention of different traditions of ideas, beliefs and ideologies that often plays an important role in attracting active participation and involvement in social movements.

For the purposes of this chapter, social movements will thus be defined as processes of political protest that mobilize human, material and cultural resources in networks linking individual actors and organizations together in pursuit of a common cause. They provide spaces in the broader culture for new forms of knowledge-making and socio-cultural learning as a central part of their activity.

The Emergence of Climate Change Knowledge

Climate change was first identified as a potentially significant public concern as one of the many aspects of an "environmental crisis" that was to lead to the emergence of environmental movements in the 1970s. Like the other social movements that grew out of the student revolts of the 1960s – those of women's liberation and anti-imperialism, in particular – the environmental movements, as they started to be called in the 1970s, were highly critical of the ways in which knowledge was produced in society, and the ways in which students were educated. Most of the active members were university and high-school students, and most of the activity was a collective learning in relation to environmental and dealing with came to be termed the environmental crisis (McCormick 1991, Jamison 2001).

The cognitive praxis of the environmental movements was based on a philosophy, or cosmology of systemic holism derived from systems theory and popularized in such books as Barry Commoner's *The Closing Circle* (1971), *A Blueprint for Survival* (The Ecologist, 1972), *Only One Earth* (Ward and Dubos 1972), as well as in the influential writings of the American ecosystem ecologists, Eugene and Howard Odum (Hagen 1992). In the environmental movements, this ecological philosophy, or ecological world-view was combined with a practical interest in appropriate, small-scale technology that was popularized in such books as *Tools for Conviviality* by Ivan Illich (1973) and *Alternative Technology and the Politics of Technical Change*, by David Dickson (1974) and practiced in new movement settings, or spaces, such as the Center for Alternative Technology in Wales, the New Alchemy Institute in the United States, and a wide range of production collectives and alternative communities (Rivers 1975). .

It was within the cognitive praxis, or knowledge-making activity, of the environmental movements of the 1960s and 1970s that climate change was first identified as a potentially significant social and political problem. In Barry Commoner's *The Closing Circle*, for instance, before the "four laws" of ecology are presented as a new political philosophy or program, the reader is introduced to the crucially important role that carbon dioxide emissions play in the so-called greenhouse effect:

> Carbon dioxide has a special effect because it is transparent to most of the sun's radiation except that in the infrared region of the spectrum. In this respect, carbon dioxide is like glass, which readily transmits visible light, but reflects infrared. This is what makes glass so useful in a greenhouse in the winter. Visible energy enters through the glass, is absorbed by the soil in the greenhouse, and then is converted to heat, which is reradiated from the soil as infrared energy. But this infrared energy, reaching the greenhouse glass, is bounced back and held within the greenhouse as heat … .Like glass, the carbon dioxide in the air that blankets the earth acts like a giant energy valve. Visible solar energy easily passes through it; reaching the earth, much of this energy is converted to heat, but the resultant infrared radiation is kept within the earth's air blanket by the heat reflection due to carbon dioxide. Thus, the higher the carbon dioxide concentration in the air, the larger the proportion of solar radiation that is retained by the earth as heat (Commoner 1971, 26–27).

Commoner's four laws of ecology – "everything is connected to everything else", "everything must go somewhere", "nature knows best", and "there is no such thing as a free lunch"- provided a set of cosmological, or world-view assumptions for the environmental movements that, in the course of the 1970s, became significant political actors in several northwestern European countries, as well as in North America. In political campaigns directed against various kinds of air and water pollution, chemicals in food and agriculture, and especially against the development of nuclear energy, environmental movement organizations, together with students and teachers at universities, learned about environmental problems (Cramer et al, 1987, Jamison et al 1990, Jamison 2001).

They also learned about alternative, "environmentally-friendly" ways to produce energy, food, and the other necessities of life that were based on an ecological worldview. Activists and academics joined together to learn how to build solar energy panels and wind energy plants, grow organic food, and try to live more ecologically, or what we today would call climate-smart, that is, find ways to develop technology that did not emit carbon dioxide and other greenhouse gases (for a contemporary overview, see Boyle and Harper, eds 1976). In the Netherlands, "science shops" were established at several universities to provide points of mediation between the academic world and the broader society, and in many other countries, the environmental movements fostered other forms of "citizen science" (Irwin 1995). The environmental and energy movements of the 1970s also inspired the formulation of new ideas about science and technology, both for the production of energy but also more generally (Commoner 1976, Lovins 1977, Nowotny and Rose, eds 1979, Capra 1982)

In Denmark, local groups in the national Organization for Renewable Energy arranged courses at many folk high schools, and created centers for renewable energy, such as the Nordic center in Thisted, which is still an in operation. In 1978, the world's then largest wind energy power plant was constructed by students at the Tvind folk high schools on the Danish west coast, not far from where VESTAS, the world's largest wind energy company, is now based (Jamison 2001). Mobilizing a Danish tradition – Poul La Cour, a folk high school physics teacher in the 19th century had been one of the first in the world to experiment systematically with wind-power generated electricity production – the Organization for Renewable Energy (or OVE, *Organisation for vedvarerende energi*) has continued to foster "grass-roots innovation" ever since. Of course, the Danish interest in wind energy was motivated, as well, by economic concerns, and, not least, by the widely felt need at the time to diminish the dependence on foreign oil, but the mobilization of an indigenous engineering tradition was important in providing valuable cognitive and cultural resources for the subsequent development of wind energy (Jamison 1978).

The Shaping of Climate Change Skepticism

In the 1980s, as the political climate in North America and northwestern Europe turned to the right, environmental politics changed character, and the making of environmental knowledge changed as well. From a social movement perspective, this right turn in politics represented a mobilization of conservative traditions, or – as they are often referred to in the United States – neo-conservative values and interests. religious and nationalist concerns were fundamental to these neo-conservative movements, which emerged, at least in part, as a kind of organized opposition to the environmental and women's movements of the 1970s and the kind of knowledge they had embodied and articulated (Helvarg 1988, McCright and Dunlap 2000, Austin 2002, Jacques et al 2008).

In many European countries, similar movements emerged at this time to oppose immigration and European integration. In Denmark, there was a strong mobilization against entrance into the European Union, and this later led to the building of the Danish People's Party which, in many ways, retains the character of a social movement even though it has become an established political party. Neo-nationalism in Europe resembles neo-conservatism in the United States, both in terms of an adherence to what might be termed a populist conception of knowledge, as well as in regard to a cosmological belief in national identity and the importance of upholding traditional values.

It is beyond the scope of this chapter to discuss these movements in any detail. However, like other social movements, they can be said to have mobilized human, material and cultural resources for purposes of political protest. As with many other recent movements, they have organized themselves through various social networks that provide opportunities for interested individuals to interact. Networks in the mass media, in radio and television, as well as in other settings, such as political action committees in the United States, have provided what can be termed the organizational dimension of these movements' cognitive praxis. Most recently, these networks have been strengthened by the Internet and the various other forms of electronic communication, but even before the Internet became widely used, a number of new think tanks and study organizations developed in both Europe and the United States to spread the ideas of neo-conservatism and neo-nationalism (McCright and Dunlap 2003; Jacques et al 2008). As such, the neo-conservative cosmology was linked to specific actions (to protest, among other things, abortion rights in the United States, and immigrant rights in Europe).

It was within the socio-cultural space carved out by these neo-conservative and neo-nationalist movements that anti-environmentalism would emerge as a political force in the course of the 1980s. Already in the debates about nuclear energy in the 1970s, a number of natural scientists, especially atomic physicists, began to challenge the forms of knowledge-making and the epistemic claims that were promulgated in the new social movements of the 1970s, particularly in the environmental and anti-nuclear movements; indeed the energy debates of the 1970s were, in large measure debates about different conceptions of science and technology. Out of those debates would later grow a defensive attitude toward what might be termed traditional science on the part of many scientists, and, more specifically, an opposition to the scientific methods and modeling techniques that would become so important in the making of climate change knowledge, in particular the complex general circulation models (Lahsen 2005, 2008).While certainly not all climate change skeptics are neo-conservatives or neo-nationalists – many skeptics are simply dubious about the kind of scientific claims that are made – it was the neo-conservative movement that provided a context for climate change skepticism to become politically significant.

Denmark is an interesting example in this regard. The rise to national and later international prominence of Bjørn Lomborg corresponds in time to the rise in political significance of neo-nationalist movements in Denmark and its political wing, the Danish People's Party (Jamison 2004). While disagreeing on many substantive issues, Lomborg and the Danish People's Party do share a common opposition to the strong emphasis that was given to "green" politics in Denmark in the 1990s and, in terms of their cognitive praxis, they share what might be termed a populist conception of science and knowledge.

The Rise of Green Business

At the same time as the anti-environmental "backlash" was taking shape in the 1980s, the environmental movement itself fragmented into a number of different organizations and institutions, both in terms of politicsnd knowledge-making (Cramer et al 1987; Jamison 2001). Green parties were formed in many countries and professional activist organizations, such as Greenpeace, grew in significance, while more broad-based, or grass-roots, organizations that had led the campaigns against nuclear energy in the 1970s tended to weaken (Eyerman and Jamison 1989). Within universities and new environmental "think tanks" such as the World Resources Institute and the Wuppertal Institute, different sorts of experts started to specialized kinds of knowledge in areas as renewable energy, organic agriculture, and eventually in relation to climate change (Jamison 1996).

As such, more professional and established forms of knowledge-making started to replace the kinds of appropriate or alternative science and technology that had been so prominent in the 1970s. Many of those who had been active in the environmental movements in the 1970s left the movement behind to make careers in universities, as well as in the wider worlds ofgovernment, media, and business. As the surrounding society became more commercial and competitive in the course of the 1990s – the result, one might say, of another social movement, namely that globalization or neo-liberalism – a good deal of green knowledge also became more commercial and competitive. Instead of learning together and cooperating with each other in projects of collective learning, many makers of green knowledge went into business (Hajer 1995, Athanasiou 1996, Hoffman 2001).

Especially in the United States, but also in many European countries, universities were encouraged to form closer ties with private companies. At first, in the 1980s, the links were primarily institutional, as offices for technology transfer and product development were established at many universities, as well as the ubiquitous science parks where companies could locate near university campuses. In the course of the 1990s, a broader process of commercialization of science took place, as new neo-liberal think tanks and a range of research institutes, often funded by private companies, started to proliferate outside of the universities (Bok 2003, Slaughter and Rhoades 2004, Hård and Jamison 2005).

For Michael Gibbons and his fellow authors of the influential book, *The New Production of Knowledge* (1994), traditional discipline-based science – what they term "mode 1" – was ever more being supplanted by approaches to science that disregard disciplinary boundaries and are directly oriented toward contexts of application. A new mode of knowledge production – so-called "mode 2" – is said to have emerged which challenges the values or norms that had previously governed the scientific enterprise. Knowledge in mode 2, we are told, has "its own distinct theoretical structures, research methods and modes of practice ...which may not be locatable on the prevailing disciplinary map" (Gibbons et al 1994: 168).

These forms of knowledge-making were supported and encouraged by new market-oriented approaches to science, technology and environmental policy that became especially important in several European countries, where social-democratic governments, often with the support of green parties, pursued policies of "ecological modernization" in the 1990s, as did the Clinton-Gore administration in the United States (Fischer and Schot, eds 1993, Mol and Sonnenfeld, eds 2000). In Germany, Great Britain, Denmark, Sweden and the Netherlands, as well as at the European Commission, ecological modernization sought to combine environmental concern with economic growth. As climate change became a more integral part of environmental politics in the 1990s, it was the market-oriented approaches that tended to dominate the international deliberations, both in Kyoto, as well as within intergovernmental administrative and scientific advisory bodies, such as the Intergovernmental Panel on Climate Change (IPCC).

From a social movement perspective, the rise of market-oriented environmentalism – or what I have termed green business (Jamison 2001) – was shaped by the broader neo-liberal movement that Leslie Sklair (1997) has characterized in social movement terms as "transnational capitalism in action". Whether we consider neo-liberalism as a social movement or as a dominant political ideology, it has certainly exerted a powerful influence on environmental politics in general and the politics of climate change in particular. Much of the knowledge-making activity within green business tends to be organized in commercial networks, with university scientists and engineers working together with companies on specific projects. There are also a number of "movement intellectuals" in the commercial media as well as in private consulting companies who serve to articulate the underlying importance of meeting the climate challenge in commercial terms. Along with Al Gore, the author and *New York Times* columnist, Thomas Friedman, have been perhaps the most publicly visible of these movement intellectuals. The "cognitive praxis" of green business exemplifies the dominant approaches of academic capitalism in the promotion of commercially-oriented technological innovation and green product development as the main "solution" to climate change.

The cosmology of green business is based on a belief in a convergence between economic growth and environmental protection, and depending on the context, it has been termed ecological modernization, eco-efficiency, corporate sustainability, or green growth. In the words of Maarten Hajer, what was central to the political discourse of ecological modernization in the 1990s was "the fundamental assumption that economic growth and the resolution of the ecological problems, can in principle, be reconciled. Hence, although some supporters may individually start from moral premises, ecological modernization basically follows a utilitarian logic: at the core of ecological modernization is the idea that pollution prevention pays" (Hajer 1995, 27). In the course of the past fifteen years, particularly in China and other Asian countries, this fundamental assumption is central to major national programs in "green growth".

In relation to climate change, one of the main proponents of market-oriented, or green business approaches has been the former U.S. vice-president Al Gore. Already in his first book, Gore (1993) combined arguments for economic growth with arguments for environmental protection in providing what he called a "new common purpose" for humanity. After the fall of the Soviet empire, the "singular will of totalitarianism" had fallen as a challenge:

> But now a new challenge – the threat to the global environment – may wrest control of our destiny away from us. Our response to this challenge must become our new central organizing principle. The service of this principle is consistent in every way with democracy and free markets (Gore 1993, 277).

In his book, Gore proposed what he then termed a "Global Marshall Plan" for saving the environment, by which he meant massive investments in renewable energy companies and in other environmentally-friendly technological developments. In the 1990s, as vice-President, Gore led the US delegation to Kyoto, where he was one of the central promoters of what has since been termed the cap-and-trade approach for dealing with climate change. After losing the 2000 election, Gore emerged as the main proponent for using market mechanisms and business ventures to respond to what he so famously called the "inconvenient truth" of climate change.

An Emergent Movement for Climate Justice

In the last few years, a new kind of political activism, often involving forms of civil disobedience and direct action, has emerged in relation to climate change and has led some observers to begin referring to a climate justice movement as a part of a broader movement for global justice.

The global justice movement has been characterized as a "movement of movements", a term coined by Naomi Klein in the wake of the anti-globalization protests of the late 1990s and which captures well the heterogeneous character of the emerging sub-movement for climate justice, as well as the broader global justice movement (Klein 2000). Both movement and sub-movement are filled with tensions and contradictions, composed as they are of a variety of groups and individuals who have begun to take political action in order to protest the quite different kinds of negative consequences that they attribute to globalization, and proposing ways of dealing with them in a more equitable, or just manner.

For the influential theorists Michael Hardt and Antonio Negri (2004), the working class or "masses" that were mobilized in the social movements of the late 19th and early 20th centuries have given way to a "multitude" of disenfranchised and disenchanted global citizens. While a multitude of voices and concerns has begun to be heard in relation to globalization and climate change, the multitude has not yet formed a shared set of beliefs that can serve as a cosmological dimension for a social movement's cognitive praxis. An awareness throughout the world is emerging about the need for a movement for climate justice, but, at least for this observer, there is little agreement as to what the movement should do and how it should organize itself. Like other social movements in their initial stages, there is as yet no real integration of the relatively abstract theorizing about global injustice voiced by theorists like Hardt and Negri with the multifaceted array of practical activities that are being carried out; there is not yet a social movement with a coherent or integrated cognitive praxis.

There are at least three different kinds of sub-movements, or networks, concerned with global justice, and which take part in various international gatherings that are sometimes said to represent the global justice movement. On the one hand, there are the parties, organizations, federations and other institutionalized legacies of the so-called "old" social movements of the 19th and early 20th centuries, the various outgrowths of populist and socialist movements that have become integral parts of the political landscape in both the global North and the global South. Issues of equality and justice for workers and farmers have been central to these movements from the outset, and in the contemporary world, they tend to base their political activity on socialist values of one denomination or another. In the emerging movement for global justice, members of various socialist organizations and parties often enter into alliances with other kinds of organizations with very different backgrounds and motivations, and as a result it has been difficult to reach agreement, or form a collective identity about particular topics such as climate change.

A second important component of the emerging movement is based on the concerns of the so-called new social movements of the 1970s, especially the movements for environmental protection, anti-imperialism and women's liberation that were so significant in the United States and northern Europe. These movements have tended to become established fixtures in the contemporary world, primarily in the form of nongovernmental organizations (NGOs) that have developed around particular issues and projects. These organizations a more business-like, professional approach to politics – they have largely become institutions rather than movements –and, much like university scientists and engineers, have become dependent on external funding for much, if not most, of their activity (cf. Jamison 1996).

In recent years, these "old" and "new" social movements have been complemented by a newer wave, or generation of activists and by groups and organizations which are often more confrontational than the older movements and more directly focused on the negative consequences of globalization, including climate change. Beginning in the 1980s, sometimes in the name of "environmental justice", these groups have often emerged in direct opposition to particular examples of global injustice, as in campaigns against the imposition of genetically-modified organisms by transnational corporations in developing countries, the construction of large infrastructural projects (dams, airports, bridges) in both developed and developing countries, and the destruction of rain forests and other biotopes in the name of economic development (Taylor ed 1995, Schlosberg 1999, Tokar ed 2001).

These are, for the most part, campaign organizations that sometimes band together in alliances in order to oppose specific cases of global injustice, but there are also a number of primarily local organizations in both the global North and global South that carry out a range of more constructive activities in relation to such areas as renewable energy, ecological housing and design and organic agriculture. In recent years, there have been attempts to arrange gatherings, where the different component parts of the global justice movement can meet and discuss their concerns, and exchange their experiences. These various "social forums", as they have come to be called, have taken place both at an international level (at world social forums, that have been held each year since 2000), as well as at more regional, national, and local levels, particularly in Europe (Fisher and Ponniah, eds 2003).

There are geographical tensions among the various component parts of the emerging global justice movement, and, as might be expected, there are thus major differences among those actively involved in regard to specific issues like climate change. Climate justice tends to mean something very different for activists in the global North than it does for activists in the global South. The very different life experiences and expectations of the participants make it difficult to develop a common understanding and shared belief system. Ideas of fairness and equity are highly dependent on contexts of history and place. As has been clear throughout the history of international climate change deliberations, there is a fundamental difference between the meaning of "climate justice" for those living in the industrialized and developed countries of the global North and those living in the industrializing and developing countries of the global South (cf. Parks and Roberts 2010).

In addition to this basic geographical conflict, there are also generational and intellectual tensions. On the one hand, the attitudes of labor organizations and social-democratic and communist parties tend to be positive toward modern science and technology and to "modernization" in general. In relation to climate change, technological development is generally seen as a central ingredient of the politics and policy of climate change, most controversially, leading to a resurgent support for the development of nuclear energy among many on the left who also purport to believe in one or another form of "climate justice" (cf. Giddens 2009). On the other hand, the institutional legacies of the new social movements of the 1970s – primarily the larger environmental NGOs – tend to see climate change exclusively as an environmental challenge and, until quite recently, have tended to disregard the social and political implications of climate change.

The task of alerting the public to the wide range of challenges that climate change raises in regard to global justice and social inequality has fallen primarily to a relatively small group of newer organizations and activists. Particularly in Africa, Asia and Latin America, groups and alliances to save rainforests, preserve biodiversity, defend the rights of indigenous peoples, and develop sustainable forms of agriculture and industry are rapidly proliferating, and some of them have begun to take part in the international climate change debate (Chawla 2009, Engler 2009, Vinthagen 2009). In North America and Europe, some of the development-oriented, and more radical environmental NGOs, such as OXFAM and Friends of the Earth, have begun to include climate change in their activities, as well. In several European climate action campaigns, and internationally through the 350.org campaign, started by the American writer Bill McKibben, a number of protest actions and climate camps have been carried out in recent years, the most noticeable being perhaps the occupations of airport runways with activists dressed as polar bears. At the Copenhagen COP15 meetings, there were several organized protest actions as well as large street demonstrations, and it is to be expected that such direct action will continue in the years to come.

There are also a growing, but still relatively small, number of cases of collaboration between academics and activists in universities and local communities in trying to deal with climate change and other environmental problems in just or equitable ways (Hess 2007, Worldwatch Institute 2010). New forms of community-based innovation and knowledge-making can be identified in local food movements around the world, as well as in a range of not-for-profit engineering projects in such areas as sustainable transport, renewable energy, and low-cost, environmentally-friendly housing. Such projects in which students and teachers from architecture and planning have designed low-cost, climate-smart housing in East Austin in cooperation with local housing suppliers and neighborhood groups show what can be done (Alley Flat Initiative 2009; cf Jamison 2009).

Unfortunately, however, such activities fall well outside of the mainstream and remain quite marginal at universities in most countries, although, in recent years, several universities in the United States have established programs in engineering for sustainable community development (Lucena et al 2010). In some of these programs there is a similar kind of institutional outreach that was so characteristic of the bridge-building activities that took place at many universities in the 1970s, but most of them have yet to achieve the influence and legitimacy that would make them significant players in the global politics of climate change. The increasing encroachment of a commercial and entrepreneurial value system at universities makes it difficult for concerns with climate justice to be given the attention they deserve in higher education.

It can only be hoped that such activities can contribute to the development of what Keck and Sikkink (1999) have termed transnational advocacy networks. The kinds of "activism beyond borders" that has been so important in political struggles in many parts of the world has often mobilized forms of local, or indigenous knowledge, which up to now have not entered into the making of climate change knowledge in a significant way. The failure of the Copenhagen meeting has led to some efforts in this regard, such as the formulation of the "people's agreement of Cochamamba" in April 2010 at the World People's Conference on Climate Change and the Rights of Mother Earth, held in Bolivia (People's Agreement 2010), and the establishment of climate justice networks in both the global North and the global South. It remains to be seen, however, whether such efforts will be able to play a significant role in the making of climate change knowledge and in establishing meaningful international agreements.

Conclusions

From a social movement perspective, climate change knowledge can be seen as a field of contention (see table 3.2.1):

Table 3.2.1   Contending Modes of Climate Change Knowledge

	Residual "Skepticism"	Dominant "Green Business"	Emergent "Climate Justice"
Broader Movement	Neo-conservative	Neo-liberal	Global Justice
Ideal of Science	Academic, disciplinary	Entrepreneurial, transdisciplinary	Cross-disciplinary
Forms of Knowledge	Traditional, personal	Context-dependent, proprietary	Collective

As I have attempted to show in this chapter, these contending modes of knowledge making are connected to broader political and social movements, with ideals of science and forms of knowledge-making. On the one hand, there is what might be termed a "residual" mode of climate change knowledge that is connected to the broader neo-conservative and neo-nationalist movements. There is, in these movements a traditional conception of science and knowledge as a kind of detached, objective truth-seeking, and it is perhaps that many scientists and engineers, and among them physicists, trained during the 1940s and 1950s, when physics enjoyed high prestige in the sciences and broader society, joined together with much more reactionary people and outright anti-environmentalists in being skeptical to the scientific claims and policy recommendations of those favoring more ambitious responses to climate change.

The dominant mode of climate change knowledge corresponds to the dominant mode of knowledge production in the world today, a "mode 2" of context-dependent transdisciplinary knowledge production, in which the traditional boundaries between science and politics and the borders between the academic and commercial worlds are increasingly transgressed (Gibbons et al 1994, Elzinga and Jamison 1995, Hård and Jamison 2005). In these contexts, science is not carried out in a disinterested and impartial fashion, but is rather funded by external interests in order to contribute directly to policy-making as well as to technological development. In relation to climate change, such policy-relevant research has been extremely important in many climate research centers, as well as in the IPCC (Shackley 1996, Hunt and Shackley 1999, Edwards and Miller, eds 2001, Grundmann 2007). Such research is often carried out in networks connecting academics, government and business in specific projects, both to provide policy advice as well as profitable "solutions" to climate change. As such, it is based on a different set of epistemic criteria than traditional, academic science, with different methods of investigation, different procedures of interpretation, and different rationales of justification and verification (Elzinga 1985, Elzinga 1996, Jasanoff and Martello, eds, 2004, Lahsen 2005, Yearley 2009).

At the same time, an entrepreneurial value system has come to challenge the traditional academic norms of science, which were influentially formulated in the 1940s by the American sociologist Robert Merton and which have long been seen by many natural and social scientists alike as core values in the scientific enterprise: communalism, universalism, distinterestedness and organized skepticism (Merton 1942). These changes in epistemology and values have led at least some, more traditionally minded scientists to treat the results of such science with skepticism (Lahsen 2008). One particularly notable example of a highly-renowned scientist who has been vocal in his skepticism is the physicist Freeman Dyson (Dawidoff 2009). The close links to business of at least some of the leading spokespersons for this kind of science, such as Al Gore, have also led to charges of conflict of interest (McGirt 2007, Broder 2009).

An emergent mode of climate change knowledge that is explicitly connected to concerns of global justice and fairness is comparatively weak at the present time, and of its future development will depend not so much on transcending disciplines as in cross-fertilizing professional and academic knowledge with local and traditional knowledge in developing a change-oriented "green knowledge" (Jamison 2010). While a number of good examples have provided sources of inspiration and mobilization in recent years, perhaps especially in relation to food and energy, there has not yet developed in this third mode of climate change knowledge a commonly shared theoretical and conceptual framework. In a world in which universities have become increasingly market-oriented and knowledge has largely been transformed into a commercial commodity, this kind of cross-disciplinary and cross-cultural knowledge making is, to put it mildly, not particularly encouraged, well-supported, or understood.

In the years to come, it will be exciting to see if the emergent movement for climate justice can develop an integrative cognitive praxis, combining ideas of global justice and citizenship with practical experiments in sustainable development. Since climate change is such an all-encompassing and multifaceted issue, it will be necessary to foster what I have termed a "hybrid imagination", mixing natural and social, local and global, academic and activist forms of knowledge in new combinations (Jamison 2008, Jamison and Mejlgaard 2010). In this regard, perhaps the efforts of the physicists-turned-environmentalists, Vandana Shiva and Fritjof Capra can provide inspiring role models. In their writings, both Shiva and Capra have tried through many years to combine disparate fields of science in an engaged and personal way, and have presented their knowledge in popular, accessible form, outside of the established academic world (e.g. Capra 1982, Capra 2002, Capra 2008, Shiva 1988, Shiva 2000, Shiva 2005). Both have also created centers for research and education and taken part in a wide range of political campaigns and struggles. Much will depend on how successful "movement intellectuals" in an emergent movement of climate justice will be in developing public spaces or sites where scientists, engineers and citizens can come together to learn from each other and bring their different kinds of knowledge into fruitful combinations.

References

Alberoni, F. (1984). *Movement and Institution*. Columbia University Press

Alley Flat Initiative, The. (2009). http://thealleyflatinitiative.org/ accessed 11/02/09

Aparacio, J and Blaser, M. (2008). The "Lettered City" and the Insurrection of Subjugated Knowledges in Latin America, in *Anthropological Quarterly*, winter.

Athanasiou, T. (1996). *Divided Planet. The Ecology of Rich and Poor*. Little Brown and Co.

Austin, A. (2002). Advancing Accumulation and Managing Its Discontents: The U.S. Antienvironmental Countermovement, in *Sociological Spectrum* 22.

Beck, U. (1992). *Risk Society. Towards a New Modernity*. Sage.

Benford, R. and Snow, D. (2000). Framing Processes and Social Movements: An Overview and an Assessment, in *American Journal of Sociology* 26

Bok, D. (2003). *Universities in the Marketplace: the commercialization of higher education*. Princeton University Press.

Boyle, G. and Harper, P. eds. (1976). *Radical Technology*. Wildwood.

Broder, J. (2009). Gore's Dual Role: Advocate and Investor, *The New York Times*, November 2.

Buechler, S. and Cylke, F. eds. (1997*). Social Movements: Perspectives and Issues*. Mayfield.

Capra, F. (1982). *The Turning Point: Science, Society and the Rising Culture*. Simon and Schuster.

Capra, F. (2002). *The Hidden Connections*. Doubleday.

Capra, F. (2008). *The Science of Leonardo*. Anchor Books.

Castells, M. (1996). *The Rise of the Network Society*. Blackwell.

Chawla, A. (2009). Climate Justice Movements Gather Strength, in Worldwatch Institute, *State of the World 2009*. Earthscan.

Commoner, B. (1971) *The Closing Circle*. Knopf.

Commoner, B. (1976). *The Poverty of Power*. Knopf.

Cramer, J, Eyerman, R and Jamison, A. (1987). The Knowledge Interests of the Environmental Movement and Its Potential for Influencing the Development of Science, in S Blume et al, eds, *The Social Direction of the Public Sciences*,. Dordrecht: Reidel

Dawidoff, N. (2009). The Civil Heretic, in The New York Times Magazine, March 29.

della Porta, D and Diani, M. (2006). *Social Movements. An Introduction*. Blackwell.

della Porta, D, ed (2007) The *Global Justice Movement: Transnational and Cross-national Perspectives*. Paradigm Publishers.

Dickson, D. (1974). *Alternative Technology and the Politics of Technical Change*. Fontana.

Ecologist, The. (1972). *A Blueprint for Survival*. Penguin Books.

Edwards, P. and Miller, C. eds. (2001). *Changing the Atmosphere: Expert Knowledge and Environmental Governance*. The MIT Press.

Elzinga, A. (1985). Research, Bureaucracy and the Drift of Epsietmic Criteria, in B Wittrock and A Elzinga, eds, *The University Research System*. Almqvist & Wiksell.

Elzinga, A. (1996). Shaping Worldwide Consensus: The Orchestration of Global Change Research, in A Elzinga and C Landström, eds, *Internationalism and Science*. Taylor Graham.

Elzinga, A and A Jamison. (1995). Changing Policy Agendas in Science and Technology in S Hasanoff, et al, eds, *Handbook of Science and Technology Studies*, Sage.

Engler, M. (2009). The Climate Justice Movement Breaks Through, at www. yesmagazine.org, accessed 25/02/10.

Escobar, A. (2008). *Territories of Difference*. Duke University Press.

Eyerman, R. and Jamison, A. (1989). Environmental Knowledge as an Organizational Weapon: The Case of Greenpeace, in *Social Science Information*, 2.

Eyerman, R. and Jamison, A. (1991) *Social Movements. A Cognitive Approach*. Polity.

Eyerman, R. and Jamison, A. (1998) *Music and Social Movements*. Cambridge University Press.

Fischer, K. and Schot, J, eds. (1993). *Environmental Strategies for Industry*. Island Press.

Fisher, W. and Ponniah, T, eds. (2003). *Another World is Possible*. Zed Books.

Garrett, R.K. (2006). Protest in an Information Society: A Review of Literature on Social Movements and New ICTs, in *Information, Communication and Society*, 9:2.

Gibbons, M. et al (1994). *The New Production of Knowledge*. Sage.

Giddens, A. (2009) *The Politics of Climate Change*. Polity.

Goodwin, J. and Jasper, J. eds (2004). *Rethinking Social Movements: Structure, Meaning and Emotion*. Lanham, MD: Rowman and Littlefield.

Goodwin, J., Jasper, J. and Polletta, F. eds (2001). *Passionate Politics: Emotions and Social Movements*. The University of Chicago Press.

Gore, Al. (1993) *Earth in the Balance*. Plume.

Grundmann, R. (2007). Climate Change and Knowledge Politics, in *Environmental Politics*, 16:3.

Habermas, J. (1984). *The Theory of Communicative Action. Volume 1*. Polity.

Habermas, J. (1987). *The Theory of Communicative Action. Volume 2*. Polity.

Hagen, J. (1992). *An Entangled Bank. The Origins of Ecosystem Ecology*. Rutgers University Press.

Hajer, M. (1995). *The Politics of Environmental Discourse*. Oxford University Press.

Hardt, M. and Negri, A. (2004) *Multitude*. Penguin.

Helvarg, D. (1988). *The War Against the Greens*. Sierra Club Books.

Hess, D. (2007). *Alternative Pathways in Science and Industry. Activism, Innovation and the Environment in an Era of Globalization*. The MIT Press.

Hoffman, A. (2001). *From Heresy to Dogma: An Institutional History of Corporate Environmentalism*. Stanford University Press.

Hulme, M. (2009). *Why We Disagree About Climate Change*. Cambridge University Press.

Hunt, J. and Shackley, S. (1999). Reconceiving Science and Policy: Academic, Fiducial and Bureaucratic Knowledge, in *Minerva*, 37.

Hård, M. and Jamison, A. (2005). *Hubris and Hybrids: A Cultural History of echnology ad Science*. Routledge.

Illich, I. (1973). *Tools for Conviviality*. Harper and Row.

Irwin, A. (1995). *Citizen Science*. Routledge.

Jacques, P., R Dunlap and M. Freeman. (2008). The organisation of denial: Conservative think tanks and environmental skepticism, in *Environmental Politics* 17.

Jamison, A. (1978). Democratizing Technology, in *Environment*, January-February.

Jamison, A. (1988). Social Movements and the Politicization of Science, in J Annerstedt and A Jamison, eds, *From Research Policy to Social Intelligence*. Macmillan.

Jamison, A. (1994). Western Science in Perspective and the Search for Alternatives, in Salomon et al, eds, *The Uncertain Quest. Science, Technology and Development*. UN University.

Jamison, A. (1996). The Shaping of the Global Environmental Agenda: The Role of Non-Governmental Organizations, in Lash et al, eds, *Risk, Environment, Modernity*. Sage

Jamison, A. (2001). *The Making of Green Knowledge. Environmental Politics and Cultural Transformation*. Cambridge University Press.

Jamison, A. (2004). Learning from Lomborg: Or Where Do Anti-Environmentalists Come From? in *Science as Culture*, June.

Jamison, A. (2006). Social Movements and Science. Cultural Appropriations of Cognitive Praxis, in *Science as Culture*, March:1.

Jamison, A. (2009). Educating Sustainable Architects. Reflections on the Alley Flat Initiative at the University of Texas. Unpublished manuscript.

Jamison, A. (2010). In Search of Green Knowledge, in S Moore, ed, *Pragmatic Sustainability*. Routledge.

Jamison, A., Eyerman, R., Cramer, J. and Læssøe, J. (1990). *The Making of the New Environmental Consciousness. A Comparative Study of the Environmental Movements in Sweden, Denmark, and the Netherlands*. Edinburgh University Press.

Jamison, A. and Eyerman, R. (1994). *Seeds of the Sixties*. University of California Press.

Jamison, A. and N. Mejlgaard. (2010). Contextualizing Nanotechnology Education: Fostering a Hybrid Imagination in Aalborg, Denmark, in *Science as Culture* 19.

Jasanoff, S. and Martello, M. eds. (2004). *Earthly Politics: Local and Global in Environmental Governance*. The MIT Press.

Keck, M. and Sikkink, K. (1999). *Activists beyond borders: advocacy networks in international politics*. Cornell University Press.

Kitschelt, H. (1986). Political Opportbnity Structures and Political Protest: Anti-Nuclear Movements in Four Democracies, in *British Journal of Political Science*, 16.

Klein, N. (2000) *No Logo: no space, no choice, no joba*. Flamingo.

Lahsen, M. (2005). Seductive Simulations: Uncertainty Distribution Around Climate Models. *Social Studies of Science* 35.

Lahsen, M. (2008). Experiences of Modernity in the Greenhouse: A cultural analysis of aphysicist "trio" supporting the backlash against global warming, in *Global Environmental Change* 18.

Lomborg, B. (2001). *The Skeptical Environmentalist*. Cambridge University Press.

Lomborg, B. (2007). *Cool It: the skeptical environmentalist's guide to global warming*. Knopf.

Lovins, A. (1977). *Soft Energy Paths*. Penguin.

Lucena, J., J. Schneider and J. Leydens. (2010). *Engineering and Sustainable Community Development*. Morgan and Claypool.

McAdam, D. and Diani, M. Eds. (2003) *Social Movements and Networks*. Oxford University Press.

McAdam, D, Tarrow, S, and Tilly, C. (2001) *Dynamics of Contention*. Cambridge University Press.

McCright, A. and Dunlap, R. (2000). Challenging Global Warming as a Social Problem: An Analysis of the Conservative Movement's Counter-Claims, in *Social Problems*, 47:4.

McCright, A. and Dunlap, R. (2003). Defeating Kyoto: The conservative movement's impact on climate change policy, in *Social Problems*, 50: 3.

McGirt, E. (2007). Al Gore's 100 Million Makeover, in *Fast Company*, July 1.

Malone, E. (2009). *Debating Climate Change*. Earthscan.

Melucci, A. (1985). The Symbolic Challenge of Contemporary Movements, in *Social Research* 52:4.

Melucci, A. (1996). *Challenging Codes*. Cambridge University Press.

Merton, R. (1942). Science and technology in a democratic society, in *Journal of Legal and Political Sociology*, 1.

Mol, A. and Sonnenfeld, D. eds. (2000). *Ecological Modernization Around the World: Perspectives and Critical Debates*. London: Frank Cass.

Moore, S. ed. (2010). *Pragmatic Sustainability*. Routledge.

Nowotny, H. and H. Rose, eds. (1979). *Counter-movements in the Sciences*. Reidel.

Parks, B. and J. T. Roberts. (2010). Climate Change, Social Theory and Justice, in *Theory, Culture and Society* 27.

People's Agreement. (2010). People's Agreement of Cochambamba, World People's Conference on Climate Change and the Rights of Mother Earth, http://pwccc.wordpress.com/2010/04/24/peoples-agreement/#more-1584, accessed June 11, 2010.

Rivers, P. (1975). *The Survivalists*. London: Eyre Methuen.

Rowell, A. (1996). *Green Backlash. Global Subversion of the Environment Movement*. Routledge.

Schlosberg, D. (1999). *Environmental Justice and the New Pluralism*. Oxford University Press.

Shackley, S. (1996). Global Climate Change and Modes of International Science and Policy, in A Elzinga and C Landström, eds, *Internationalism and Science*. Taylor Graham.

Shiva, V. (1988). *Staying Alive. Women, Ecology and Development*. Zed Books.

Shiva, V. (2000). *Stolen Harvest. The Hijacking of the Global Food Supply*. South End Press.

Shiva, V. (2005). *Earth Democracy. Justice, Sustainability and Peace*. South End Press.

Sklair, L. (1997). Social Movements for Global Capitalism: the transnational capitalist class in action, in *Review of International Political Economy*, 4:3.

Slaughter, S. and Rhoades, G. (2004). *Academic Capitalism and the New Economy*. Johns Hopkins University Press.

Smelser, N. (1962). *Theory of Collective Behavior*. Routledge and Kegan Paul.

Snow, D. and Benford, R. (1988). Ideology, Frame Resonance, and Participant Mobilization, in *International Social Movement Research* 1.

Taylor, B. ed. (1995). *Ecological Resistance Movements. The Global Emergence of Radical and Popular Environmentalism*. Albany: SUNY Press.

Tokar, B. ed. (2001). *Redesigning Life? The Worldwide Challenge to Genetic Engineering*. Zed Books.

Touraine, A. (1981). *The Voice and the Eye: An Analysis of Social Movements*. Cambridge University Press.

Touraine, A. (1988). *Return of the Actor*. University of Minnesota Press.

Vinthagen, S. (2009). The Birth of a Global Climate Justice Movement, at http://resistancestudies.org, accessed 25/02/10.

Ward, B. and Dubos, R. (1972). *Only One Earth*. Penguin.

Williams, R. (1977). *Marxism and Literature*. Oxford University Press.

Worldwatch Institute. (2010). *State of the World 2010*. Earthscan.

Yearley, S. (2009). Sociology and Climate Change after Kyoto. What Roles for Social Science in Understanding Climate Change?, in *Current Sociology*, 57:3.

The Matter of Climate Change: What It Is and How to Be Concerned With It

Gert Goeminne

Introduction

The current environmental crisis is commonly understood as a human failure to find a sustainable way of interaction between culture and nature. In recent years, the Copenhagen climate summit (December 2009) has accordingly been cast as a failure resulting from the political incapacity to stand up for a stable climate. Rather than endorsing this viewpoint, this chapter will depart from the argument that the very way it is formulated is symptomatic of the key cause of the climate change impasse, namely the lack of conceptual and political space required to think of climate change as a 'nature-culture' hybrid and deal with it accordingly. The general aim of my research project of which this chapter makes part, is to conceive of the environment as a genuinely political concept, open to struggle and contestation, in this way constituting an essential component of social change. The underlying starting point is that the current environmental policy approach of 'rational decision making' inherently rejects a plurality of socio-environmental visions, invoking a conception of both science and politics that is characterized by rationalism and universalism. As opposed to this essentialist belief in a rationally attainable consensus without exclusion, this chapter puts forward the constitutive role of exclusion in the emergence of both natural and social order. Invoking Bruno Latour's constructivist approach, I will elaborate on a practice-inspired account of climate science in which the construction –or rather the 'composition'- of a matter of fact such as anthropogenic climate change necessarily implies a significant and non-trivial differentiation between what is taken into account and what not, that is to say between internalities and externalities. This will eventually enable us to conceive of climate politics as being constituted by a struggle about what to be concerned about, a struggle that is ultimately based on the idea that every composition, including a scientific one, excludes and differentiates.

Prologue: Apocalypse Now?

"Copenhagen or the apocalypse", that is how one might describe the atmosphere that ruled the build-up to the UN climate summit in Copenhagen, December

2009. As argued by most, if not all, climate activists, the case was cut and dried: not only would the war on climate change be settled in Copenhagen, but also the fate of humanity. Climate change science as voiced by the United Nations' Intergovernmental Panel on Climate Change (IPCC) is clear on this: a further increase of CO_2 in the earth's atmosphere leads to catastrophic environmental and human consequences and in order to avoid these the growth in worldwide CO_2-emissions needs to be halted in the next 10 years. Unless we fancied ending up in a doomsday scenario, the only possible outcome left for the Copenhagen climate negotiations was a legally binding agreement setting the necessary per country emission goals (target and timeframe). It comes as no surprise then that Copenhagen has been depicted as a political failure, rich and poor countries blaming each other for the weak outcome: the UN climate summit only reached a weak outline of a global agreement recognizing the scientific case for keeping temperature rises to no more than 2 degrees Celsius but does not contain commitments to emission reductions to achieve that goal. The conclusion seems clear: climate politics has failed; the crack of doom is drawing near. Or not?

Well, it depends on what is meant by 'climate politics'. In the following I want to argue that the fact that the Copenhagen climate summit has indeed been cast as a political failure to find a sustainable way of interaction between nature and culture is symptomatic of the key cause of the climate change impasse, namely the lack of conceptual and political space required to think of climate change as a 'nature-culture' hybrid and deal with it accordingly. Climate change, while from the perspective of natural science is primarily a matter of CO_2-concentrations is also, and arguably as much, a matter of the machinations of the automobile industry, the politics of the Kyoto Protocol, the evolutions in climate modeling, the inequities in North-South trade relations, the imminent flooding of the Maldives et cetera.

For more than 15 years however, the UN-mandated IPCC has been trying to frame the climate issue as a scientific puzzle, while their scenario calculations, in which the leading part is played by CO_2, enjoy enormous political authority[1]. The result of all this is that climate change, which as a nature-culture hybrid is first and foremost a political issue questioning the organization of our society, is

---

1   In this respect, it is interesting to have a closer look at how the IPCC, now generally considered as a neutral assessment panel, actually came into existence (See Agrawala (1998) for a good overview). With global concerns about climate change coming in the wake of the ozone depletion issue, first attempts to construct a global climate policy were 'modeled' on the straightforward science-driven ad-hoc approach that led to the very succesful Vienna Convention on Ozone (1985). However, this approach failed as greenhouse gas emissions, closely connected to the use of fossil fuels, are much more socio-economically 'pervasive' than ozone-depleting substances. The increasing awareness of this political 'sensitivity' of the climate change issue eventually led to the installation of the IPCC as a kind of science-policy mediator in order to ease political disagreement by precluding immediate action ('buying time') and strengthening scientific consensus through international participation ('institutionalizing the consensus').

reduced to the technocratic management of CO_2-emissions: human action is no longer weighed against the question of the good life but against the question of how much CO_2 we emit in this way. In this widely embraced policy approach of 'rational decision making', science is playing a perfidious role: to the extent that it gives environmentalism a voice to speak up, the latter's political potential for change is immediately confined to a science-based quest for efficiency.

All this goes to show the need for a renewed understanding of climate politics that takes the question of the good life at heart while at the same time acknowledging our technoscientifically conditioned situation. In the following, I will present a first attempt to do so in three consecutive steps. Connecting a constructivist understanding of science with an antagonistic conception of the political, I aim to lay down the contours of an environmental politics of composition that is constituted by a struggle over what to be concerned about, a struggle that is ultimately based on the idea that every composition, including a scientific one, excludes and differentiates. In a first chapter, however, I wish to further elaborate my critique of the current 'rational decision making' approach to climate policy as it is conducted within the contours of the UN climate treaty.

Rational Decision Making: the Perfidious Alliance between Science and Neoliberal Climate Policy

In the political arena of our Western techno-scientific culture, societal issues and in particular environmental issues are predominantly staged in a scientific way. Environmental problems, which are more than often generated by the products of science and technology – let us not disregard this[2], are indifferently framed in scientific terms (CO_2 concentrations, Sv dose equivalents,...) which inevitably gives rise to a quest for scientific solutions (carbon tax, dose limits,...). Whether the issue is climate change or nuclear energy, the resolution is sought in ever more science (See e.g. Sarewitz (2004) and Goeminne (2011a)). This not only leads to the raging environmental controversies between believers and non-believers we are experiencing today regarding the validity of the answers science provides. More fundamentally, this one-dimensional discussion leaves the scientific questions, i.e. the way science frames environmental problems and solutions, unquestioned.

Underlying this one-dimensional approach is a strict ontological division between the non-human object and the human subject with a corresponding epistemology of represented facts and values that Bruno Latour has called the 'Modern Constitution' (Latour 1993: 13–14). In our modern culture, Latour

---

2   This not to say that I qualify science and technology as inherently bad; it rather points towards the detrimental environmental and social impacts of the specific way science and technology are embedded in our neo-liberal order. In the following, I will therefore not argue to put science and technology as such in question but rather the societal order in which they are embedded.

argues, hard scientific facts and technological instruments are systematically separated from soft values, choices and responsibilities. In such a two-tiered world of facts and values, the classical political representation of humans can only appear as an incomplete and contingent process compared to the authoritarian status enjoyed by scientists when speaking about non-humans. As an immediate effect of this, political judgment increasingly submits itself to the unequivocality of scientific statements and efficiency of technological developments. 'Rational decision making' is the now commonly used term to characterize a politics that gathers around undisputable scientific facts and that embraces technocratic management of remaining scientific uncertainties and technological risks as its core business. The politics of climate change policy, as it is conducted within the intertwined institutional and scientific contours of the UN, serves as a preeminent example here. Although going almost unnoticed, actual CO2-based climate policy paralyzes the political struggle in the sense that it does not allow contesting and discussing alternative visions of society. Rather than questioning the reigning socio-economic order by imagining and formulating alternatives, a one-sided CO2 reduction strategy is organized within the existing neo-liberal order. In this respect, it is indeed telling that CO_2 is now a tradable commodity. Swyngedouw (2010) draws a similar conclusion when he states that "the undisputed matters of fact (expect by a small number of maverick scientists) are, without proper political intermediation, translated into matters of concern." And so it happens that the political question of which society we want gets reduced to the question of which 'low carbon economy' we want, for there is apparently no doubt it needs to be a (free) market economy. Along this line, the political debate between alternative visions of society is ultimately reduced to a consensus exercise making everybody toe the neo-liberal line unless he or she wants to be held responsible for the end of humanity. The only kind of disagreement that is tolerated is limited to the choice between particular technologies and market mechanisms, a disagreement that is typically dissolved through a process of negotiation of (economic) interests (Swyngedouw 2010). In my view, it is only within such a technocratic logic that a technology such as 'carbon capture and storage' can appear as a viable alternative to save our planet. A similar reasoning backs the revival of the nuclear option: it satisfies our hunger for energy and it is CO_2-neutral, so what is the problem? Apparently, nobody seems to bother what kind of better society could be associated with such technological options. And so it also happens that the development and promotion of the electric car is now widely embraced as one of the spearheads of climate policy. Cynically enough, one might argue, innovation has become the new credo since innovation, embedded in a market logic of competitiveness, ultimately stands for the way we can continue as before, producing and consuming in order to keep economic growth on pace. Developing an engaging vision on the future where the car has lost its 'raison d'être' does not belong to the kind of innovation we need.

For the sake of uncovering the real politics of such consensual climate policy, it is important to see that the choice for a rational decision-making approach, although being put forward as such, is anything but a neutral, 'evident' one. It is indeed very revealing that sceptics of the consensual UN-process, strategic as their purposes of contesting the IPCC-consensus may be, are indifferently depicted as outlaws sometimes accusing them of a 'crime against humanity' or threatening them with court action[3]. However benign the intentions of such public convictions may be, they show the really divisive and non-innocent politics of consensual climate policy: it separates those who subscribe to the neo-liberal order from those who do not. At the heart of this neglect for the very divisive character of actual climate policy lays a lack of critical attitude vis-à-vis the very way science frames its issues. The latter became very clear during the so-called 'climategate' scandal that erupted in late November 2009. The ensuing discussion on the status of climate science took place in a one-dimensional discursive space, spanned by the terms false and true. Moving beyond truth claims that indifferently pertain to the answers of science, the next chapter will focus on the very way science frames its questions thereby drawing on Bruno Latour's constructivist account of science as a major source of inspiration.

Rethinking Science: When Matters of Fact Become Matters of Concern

In a first step towards a rethinking of climate politics, I thus argue for a closer examination of scientific practice. This has also been Latour's initial focus. Through ethnographic studies on scientists working in their laboratories and field sites, Latour has indeed taken his entry through what he himself calls "the back door of science in the making, not through the more grandiose entrance of ready made science" (Latour 1987: 4). Over the past few decades, this kind of research has figured centrally in the broad field of 'Science and Technology Studies' (STS) taking down the unworldly image of science as a truth-speaking device and replacing it with a practice-inspired account of science as culture (See e.g. Pickering (1992)). In arguing that matters of fact are always also matters of a particular concern, Latour has explicitly thematized the inherently human, that is to say, 'concerned' character of scientific practice (Latour 2004a, 2004b). Within the context of its construction, scientific knowledge aims to fulfil a certain function, and the choice of that function depends on the scientist's concern: What kind of knowledge is aimed at? What is it supposed to account for and to take into account? Beyond construction and representation, I want to appeal to the notion of 'composition' to convey the idea that a scientific fact is not chosen or given;

---

3   One of Belgian's most prominent climate publicists, Peter Tom Jones, recently stated in one of Belgium's leading newspapers that "climate skepticism is a crime against humanity" (De Standaard, 8 December 2009). Belgian climate scientist and IPCC vice-president Jean-Pascal van Ypersele has repeatedly argued in the Belgian press that "we should think seriously about bringing them (climate skeptics) to court" (e.g. De Morgen, 4 November 2006).

rather, it is 'concernfully' composed as a 'matter of concern' in relation to what is considered to be the issue at stake (Goeminne and François 2010).

Wynne is clearly on this compositionist line when he argues that "woven into the disciplined scientific attempt to understand what nature is saying to us about changing climate processes, and human responsibilities for them, are always ancillary but constitutive concerns and commitments." (Wynne 2010) In his latest book *A Vast Machine*, Edwards convincingly demonstrates that what it means to observe 'global climate change' is intrinsically intertwined with the concerns that have guided climate modellers in their daily practices (Edwards 2010). Conceptualizing such a thing as the global climate is indeed preconditioned on a concern for homogeneity, which Edwards shows to have played a crucial role in constructing reliable climate knowledge out of a vast array of disparate information by means of modelling techniques. Demeritt, on the other hand, has argued that a concern for simulation and prediction constitutes another crucial ingredient of the history of climate modelling by showing how atmospheric scientists in the 1970's deliberately tapped into growing public concerns about human impacts on the environment to secure funding for basic modelling research (Demeritt 2001). Indeed, the only way to demonstrate the anthropogenic character of climate change is to simulate what would have happened without humans adding greenhouse gasses (GHGs) to the atmosphere.

Climate change 'matters of fact' such as CO_2-concentrations are thus always also 'matters of concern', that is to say the results of a process of composition governed by a concern for –amongst other particular concerns- homogeneity, simulation and prediction. Beyond the typical STS-style recognition of the non-neutral, ambivalent character of science, I want to illuminate the constitutive role of concern in the emergence of scientific knowledge. Conceiving climate modeling, for instance, as a concernful work of composition along the lines sketched above indeed suggests that particular concerns at the upstream end play a crucial role in shaping the final form and content of the knowledge emerging at the downstream end.

Rethinking Politics: The Antagonistic Dimension of Scientific Compositions

In a second step, I now want to propose a political reading of my 'compositionist' epistemology by connecting the constitutive role it attributes to concern with an antagonistic conception of politics. Within recent political theory, a fundamental critique has been formulated vis-à-vis the conception of the political holding sway in a great deal of contemporary democratic thinking which is characterized by rationalism and universalism. Invoking the term 'post-politics', Mouffe (2005) and Rancière (1998), amongst others, lament the evacuation of antagonistic notions such as exclusion, adversary and contestation from the political sphere, which reduces politics to a mere instrumental conception focused on the consensual administration of environmental, economic and other domains. Contrary to such an instrumental view, they argue that the political should be conceived as an ontological dimension that determines our very human condition. Indeed, according to Mouffe, human society is essentially political because, first of all, "the need for collective identifications will never disappear since it is constitutive of the mode of existence of human beings" (Mouffe 2005: 28) and, secondly, as "in the field of collective identities, we are always dealing with the creation of a 'we' which can exist only by the demarcation of a 'they'." (Mouffe 2005: 15)

At this point, I propose to expand this non-essentialist thesis from the social to the natural sphere, blurring the boundaries between both. I want do so by arguing that the construction of objectivity more generally (identity, knowledge...) is relational and that its condition of existence is the affirmation of an exclusion. This constitutes the core of the argument I want to articulate here, bringing together my compositionist STS-account of science with an antagonistic conception of the political: science, conceived as a situated, concernful work of composition, is necessarily political, separating internalities from externalities. In the context of climate modelling, Demeritt (2001) has argued for instance that it is only by excluding the messy social relations that drive greenhouse gas emissions and by focusing narrowly on their universal physical properties that atmospheric scientists, concerned as they were about homogeneity and predictability, have been able to compose the issue of climate change. At this point, it should already be clear that a differing interpretation of the 'situatedness of science', that is to say the awareness that science is a human and therefore necessarily perspectival and value-loaded endeavour, lies at the crux of differentiating my 'science is political' claim from most of the STS-based claims about the non-neutrality of science. The latter -negatively- understand the situatedness of science as resulting in a restricted, suboptimal knowledge of the problem at stake that becomes the more 'contaminated' by uncertainty and value-loadings, the more 'situated' the issues get. My claim, however, positively understands the situated, perspectival character of science as being truly constitutive of the knowledge composed and as inevitably resulting in a division between what has been taken into account in the composition and what has not.

Rethinking Climate Politics as an Unending Work of Composition

In connecting Latour's constructivist account of science with an antagonistic conception of the political through the notion of composition, I have tried to clear the way for a simultaneous rethinking of the cultural and the natural spheres as one political order through the constitutive role of concern. Shifting focus in this way from a dispute over matters of fact in terms of true and false to a struggle for matters of concern in terms of internalities and externalities, a conceptual space has been disclosed that allows the environment to be composed as a nature-culture hybrid. Whereas 'rational decision making', invoking the universalism of an inert, unmediated nature, addresses itself to an undifferentiated universal 'we', a constructivist-inspired political account of knowledge recognizes the complex ways knowledge producing practices configure and reconfigure networks of relations in ways that both enable and constrain our capacity to know and to act. Invoking my claim that science is political, I thus tend to argue that UN's science-policy approach of 'translating' scientific into political consensus is fundamentally flawed as it is preconditioned on the alleged non-exclusive character of science. It has indeed already been argued that the attempt to create a carbon-trading scheme predicated upon the scientific universality of GHG emissions, in effect erases the historical origins of the emissions, excluding the difference between luxury and subsistence emissions from further policy debate (Agarwal et al. 1991).

Lost in the translation from science to policy, the concernful work of composition that goes into the construction of a matter of fact is obscured in consensual decision making leaving policy nothing but externalities to be managed in a technocratic way. Understanding the task of raising and addressing matters of concern as a work of composition, however, is the true political heritage of constructivism, conceiving politics as a struggle for who and what is to be taken in to account (Goeminne and François 2010). Such a struggle presupposes the openness towards "divergent, conflicting and alternative trajectories of future socio-environmental possibilities", the composition of which constitutes the very political work that needs to be done (Swyngedouw 2010). Also central to this approach is the awareness that such a work of composition is an unending task, as exclusion and antagonism are at the same time its condition of possibility and the condition of impossibility of its full realization (Mouffe, 1993).

This is convincingly illustrated by the way in which genuine concerns of local communities in developing countries, typically framed in terms of environmental justice, are systematically repudiated in international debates on environmental issues such as climate change (Martinez-Alier, 2002). Rather than originating in a scientific context, their struggle for environmental justice is motivated by a day-to-day confrontation with the externalities of an energy and resource guzzling Western development pattern including the clearance of rainforest land for the sake of animal fodder production or the environmental and social impacts of oil and uranium exploitation. Along the lines sketched above, we can now characterize their struggle as genuinely political, understanding it in terms of a collective identification around the constitutive outside of the neo-liberal economic paradigm. Indeed, the truly interesting point about these externalities of the Western economic paradigm is that they have the potential to fundamentally question its neo-liberal foundations, provided the necessary political space is rendered available (Goeminne and Paredis forthcoming). However, within the post-political contours of technocratic consensual policy-making, where externalities are at best internalized in terms of market corrections, environmental justice's political demand for these externalities to be taken into account as part of a different socio-environmental composition can only appear as radical opposition, so that it is sidelined. It is indeed symptomatic of their incapacity to think of 'the excluded', that consensual approaches, notwithstanding their emphasis on inclusion and participation, typically turn out to be exclusive in a dogmatic way: everybody is included as long as they play the consensual rules of the game.

Based on the same line of reasoning along which the struggle for environmental justice can be understood as a paradigmatic case of the politics of composition I envisage, I propose to acknowledge the so-called Copenhagen failure as a genuinely political result. Partly due to the unbending posture adopted by some developing countries, the consensual approach of UN climate policy that is imposed under the threat of an apocalyptic CO_2-story has encountered its own post-political limits at Copenhagen[4]. The time is right to revive the climate and by extension the environment as a matter of genuine political concern, open to struggle and contestation between alternative visions of society, in this way constituting an essential component of social change. Conceived as a matter of concern, climate change cannot be moulded into scientific criteria to be met by some low carbon economy, nor is it reducible to mere personal opinion or preference. Composing the matter of climate change as a matter of concern rather entails a political struggle for whom and what to care about and calls upon everyone and everything in rethinking and recomposing our society.

Epilogue: Towards a Politics of the Imaginable

This eventually brings me to my plea for moving from a 'politics of the knowable' to a 'politics of the imaginable' (Goeminne 2011a). Politics of the knowable, of which I have argued UN climate policy is a paradigmatic example, is founded upon 'what can be known'. Within the contours of such a politics, sustainable futures are typically conceived of in terms of techno-scientific parameters such as CO_2-emissions. The latter then set the boundaries of societal debate which subject is now limited to the technologies and market mechanisms that should be deployed to reach these criteria. In such a scientifically preconditioned context, societal debate is easily reduced to a negotiation of technological risks and (economic) interests, neglecting the very concrete 'forms of life' and socio-material practices that should eventually constitute a sustainable future. A 'politics of the imaginable' however, is founded upon 'what can be imagined', conceiving of alternative 'forms of life' that can be contested and discussed in terms of the good life.

To illustrate how such a 'politics of the imaginable' may differ from the reigning 'politics of the knowable', I revert once more to the widespread enthusiasm surrounding the development of the electric car as being one of the spearheads of the transition to a low carbon economy. Here, the electric car is typically promoted

---

4   The actual dynamics of the Copenhagen negotiations are of course too complex to attribute its impasse solely to the unbending posture adopted by the developing countries. Also, the diversity in the group of countries that make up the official delegation speaking for developing countries, i.e. G77 + China, makes it impossible to univocally attribute them the 'environmental justice' concerns mentioned.

within a matter-of-fact-based discourse revolving around carbon efficiency; a clear case of 'politics of the knowable'. In this respect, it is noteworthy that the public debate mostly focuses on those aspects in which the electric version still falls short when compared to the classical gasoline and diesel cars: a smaller range, limited acceleration power and a lower top speed. Removing these so-called shortcomings is then put forward as the evident challenges for an innovative economy. Apparently, innovation has to bring more of the same. However in thinking about this from a 'form of life' perspective, the electric car could be 'imagined' within a new kind of rationality, turning these alleged shortcomings into societal blessings. Not only could its lower top speed and acceleration power be embraced as a way to establish a safer and less stressful traffic situation, its limited range could moreover be seized upon to question the desirability of an unlimited automobility in terms of 'the good life' and could constitute a concrete starting point in conceiving alternative ways of organizing our daily life[5].

This example further shows that my notion of a 'politics of the imaginable' does not have to be associated with revolutionary ideas, calling for the radically new. Rather, a 'politics of the imaginable' takes its point of departure in what is already there, thereby arguing for a recomposition of already existing elements within a new logic beyond the allegedly neutral criterion of efficiency. Once again, it is important to see that a full appreciation of the technoscientific embeddedness of our contemporary lives implies due attention for the ways in which science and technology co-shape human practices in non-neutral ways (Goeminne 2011b). It is only by seeing how human existence is always already interweaved with science and technology within existing 'forms of life' and by imagining new ways of interweaving them within new 'forms of life', that we can begin to conceive of a 'politics of the imaginable' that is capable of addressing the question of the good life in our contemporary world.

References

Agarwal, A. et al. 1991. *Global Warming in an Unequal World*. New Delhi: Centre for Science and Environment.

Agrawala, S. 1998. Context and early origins of the Intergovernmental Panel on Climate Change. *Climatic Change 39(4)*, 605–620.

Demeritt, D. 2001. The Construction of Global Warming and the Politics of Science. *Annals of the Association of American Geographers* 91(2): 307–337.

Edwards, P. 2010. *A Vast Machine*. Cambridge (MA) & London: The MIT Press.

---

5   In this respect, this approach bears resemblances with Andrew Feenberg's theory of demcoratic rationalization, the latter focussing on transforming technological regimes through users re-appropriating a technological artefact within a new regime (Feenberg 2002).

Feenberg, A. 2002. *Transforming Technology. A critical theory revisited.* Oxford/ New York: Oxford University Press.

Goeminne, G. 2010. Once upon a time I was a nuclear physicist. *Perspectives on Science* 19 (1), 1–31.

Goeminne, G. 2011a. Has science ever been normal? On the need and impossibility of a sustainability science. *Futures* 43 (6) (in press).

Goeminne, G. 2011b. Postphenomenology and the politics of sustainable technology. *Foundations of Science* 16 (2–3) 173–194.

Goeminne, G. and Paredis, E. (forthcoming) The concept of ecological debt: challenging established science-policy frameworks in the transition to sustainable development. In: Techera, E. (ed.) *Frontiers of Environment and Citizenship*, Oxford: Interdisciplinary Press (in press).

Latour, B. 1987. *Science in Action: How to Follow Scientists and Engineers through Society.* Cambridge (MA): Harvard University Press.

Latour, B. 1993. *We Have Never Been Modern.* Cambridge (MA): Harvard University Press.

Latour, B. 2004a. Why has Critique Run out of Steam? From Matters of Fact to Matters of Concern. *Critical Inquiry*, 30:2, 225–248.

Latour, B. 2004b. *Politics of Nature: How to Bring the Sciences into Democracy.* Cambridge (MA): Harvard University Press.

Martinez-Alier, J. 2002 *The environmentalism of the poor. A study of ecological conflicts and valuation.* Cheltenham: Edgar Elgar.

Mouffe, C. 1993 *The Return of the Political.* London: Verso.

Mouffe, C. 2005. *On the Political. Thinking in Action.* London: Routledge.

Pickering, A. 1992. *Science as Practice and Culture.* Chicago (IL): University of Chicago Press.

Rancière, J. .1998. *Disagreement.* Minneapolis: University of Minnesota Press.

Sarewitz, D. 2004. How science makes environmental controversies worse. *Environmental Science & Policy* 7, 385–403.

Swyngedouw, E. .2007. Impossible 'sustainability' and the post-political condition, in Krueger, R. and Gibbs, D. (eds.) *The Sustainable Development Paradox.* New York: Guilford Press 13–40.

Swyngedouw, E. 2010. Apocalypse forever? Post-political populism and the spectre of climate change. *Theory, Culture & Society* 27(2–3), 213–32.

Wynne, B. 2010. Strange weather, again: climate science as political art. *Theory, Culture & Society* 27(2–3), 289–305.

Chapter 3.4

Education, Active Citizenship and Applied Social Intelligence: some Democratic Tools to Meet the Threat of Climate Change[1]

Joshua Forstenzer

As of this writing, the rapid deterioration of the environment on which we depend for our survival remains for the most part unimpeded. After two decades of international deliberation, we are yet to have negotiated a binding agreement between all national authorities effectively regulating greenhouse gas (GHG) emissions. All the while, climate scientists have continued to release growingly pessimistic predictions about the likely outcomes of climate change and about the prospect of slowing the warming of our environment (see, for example, Homer-Dixon 2007). Out of a number of foreseeable consequences of this warming, the following stand out as particularly worrying: rising sea levels, chronic droughts, multiplication of violent storms and hurricanes, unpredictable floods, large forest fires, rapid reduction in biodiversity (IPCC 2007).

The net effect of such developments on human life is difficult to appraise with much precision, but the likely outcomes of such changes are largely thought to include: mass migrations,[2] drastic reduction in access to clean drinking water, potentially permanent global food shortages, loss of and damage to vital infrastructure (hospitals, schools, roads, means of communication, etc.), widespread scarcity of basic medicines and the multiplication of epidemics (Bates, Kudzewicz, Wu and Palutikof 2008: 33–76, Brown 2008: 1–129, IPCC 2007).

The principal institutional consequence of such developments is expected to be major political destabilisation (Brown 2009). In particular, it is thought that brewing international tensions would likely transform into large-scale military conflicts (Homer-Dixon 1999; Stern 2007: 56–91). According to Lester R. Brown

---

1   I would like to thank Daniel De Arriba and Bijal Shah for their moral support during the drafting of this chapter. I would also like to thank Megan Kime, Abdi-Aziz Suliman and the anonymous referee for their helpful comments on previous drafts of this chapter. Finally, I would like to thank Sasha Maria Christiansen for giving me a reason to write it.

2   In fact, some say that migration caused by climate change has already started, see G. Monbiot, 'Climate change displacement has begun – but hardly anyone has noticed', Blog, *The Guardian*, 08/05/2009: http://www.guardian.co.uk/environment/georgemonbiot/2009/may/07/monbiot-climate-change-evacuation (16/05/09)

(2008: 14), the most dangerous political effect of climate change is the likely proliferation of failed states. Under the mounting pressure caused by hikes in food and energy prices, many of the less stable national governments of the world are likely to "lose control of part or all of their territory [... and thus fail to] ensure the personal security of their people."

Moreover, the effects of a failed state are usually also felt beyond its borders. According to Brown (2008: 14–15), the risk of armed conflict (most likely in the form of civil war) spreading to adjacent countries, along with the likely displacement of population and the threat of uncontrolled epidemics are likely to cause serious political destabilisation across any region harbouring but a single failed state. Not only that, but in such an uncertain environment, the multiplication of large scale conflicts across the globe is made all the more likely by: first, the limited desire of any country to shelter endangered foreign citizens (already today, immigration policies and camps of displaced populations are sources of serious political tension, and we can only expect the political fallout to increase in magnitude as the size of displaced populations increases); and second, the desire to control strategically important scarce resources –such as, clean water, food and energy– would likely drive nations to pre-emptively assault competitors or directly take control of foreign assets. Thus, the political cost of climate change is likely to be the proliferation of violent conflict.

Such a realisation can only heighten the sense of urgency bestowed upon us by the knowledge that our unabated GHG emissions continues to push us ever faster toward several tipping points, after which serious climatic changes of the sort discussed above are predicted to be all but inevitable. Meaningful action is desperately needed. And yet, after the political fiasco of the COP15 in December 2009, it appears ever more likely that the binding global political agreement necessary to address the problem structurally (i.e. by severely limiting GHG emissions by means of the law) will not come in time to stop the first dominoes from falling.

In my opinion, it is the realisation that we face this rather depressing outlook which has precipitated a broadening of the discussion about available means to address problems related to climate change. The most obvious product of this shift is the growing consideration offered to potentially hazardous geo-engineering strategies designed to slow – and/or mitigate the effects of – the immediate warming of the globe (see, for example, Crutzen 2006; Cicerone 2006). In contrast, the pursuit of cultural means to cope with climate change represents a potentially less hazardous, yet plausibly fruitful, avenue in our quest to prevent or minimise the human suffering likely to result from unimpeded climate change.

In this chapter, I will argue that John Dewey's philosophical conception of democracy can play a significant role in helping us come to grips with how potentially successful cultural responses to climate change might come about in a democratic context. As an immediate response, one might be sceptical of the project of this chapter:

"It is not obvious that philosophy has much of a role to play in helping us understand how we might conceive of such cultural change. Philosophers are in the business of pursuing abstract truths, not studying the real world and making predictions about it. Why not leave policy experts, technologists, engineers, sociologists, economists and psychologists to deal with this most concrete and pressing problem?"

It is to respond to this worry that I will begin with a section dealing with the particular manner in which philosophy might be relevant to the issue of developing cultural responses to climate change. I will then offer something of an argument in favour of adopting John Dewey's conception of democracy as a potential vision of how cultural change could be meaningfully brought about in a liberal democracy.

Philosophy, Culture and Politics: Democracy or Elitism?

Traditionally, the relationship between philosophy and culture (not to be confused with high culture) is defined by tension and conflict. To offer something of a caricature: the term 'culture' is usually thought to encompass the pre-reflective habits and beliefs of mere lay people; whereas the cold logic of philosophy painfully brings such habits and beliefs into doubt. And, indeed, when we are first introduced to philosophy, we are customarily told that Socrates spent his days wandering the *agora*, in the 5th century BC, questioning Athenians about their conduct and beliefs, challenging the dogmas and cultural norms of his time, calling into question faith in the established ways of thinking and doing things. Famously, this attempt to engage with Athenian culture cost Socrates his life (Plato 1993: 29–68).

But according to Socrates' most illustrious student, Plato, this challenge to the establishment through philosophical questioning was not the mere product of an eccentric old man's curious mind. Much to the contrary, for Plato, the rigorous scrutiny of our established ideas, beliefs and habits was the only way to overcome the ignorance imposed by the rule of mere opinion, which all too often serves as the pillar of human culture. Instead, philosophy is thought to offer a more secure and objective kind of knowledge, rooted in the certainty of reason as opposed to mere tradition. On this view, philosophy can thus offer a systematic appraisal of culture and emit recommendations to render it more rational (see Kraut 2009).

This conception of the relationship between philosophy and culture is most visible in the long tradition of political philosophy, where many philosophers have expounded their diverse visions of the good society. As a result, it should be evident that philosophy has long participated in the task of cultural and political critique. And, in fact, when we hear our political, intellectual and even artistic leaders calling for a new self-conception, a new philosophy centred on the rich and complex relationships we entertain with each other and with the environment, it would appear that philosophy has already begun to play a significant part in contemporary cultural attempts to deal with climate change.

However, an informed reader might point out that, historically, philosophy has mostly failed in its attempts to morally improve human culture. As beautiful as they are, the ideals of philosophers, for the last two millennia, have mostly gone unnoticed by the many, and served as justification for despotic regimes when taken seriously by the few. Thus, in the present context it is hard to hear exhortations for a new 'eco-philosophy' as anything more than an empty hope, a childish incantation for a *deus ex machina,* that takes us away from the very real task of changing our methods of production and consumption. We must thus confront the question: how can philosophy, a highly theoretical endeavour, possibly contribute to the practical task of meeting the challenge of climate change?

I contend that philosophy can help principally by identifying relevant cultural and political tools we might want to develop and use to confront our collective problems within a democratic context. Crucially, I think we need to address this concern forcefully and meaningfully because, as often happens when the security of a people is severely threatened, the ideological appeal of undemocratic means of solving our problems is surging. And, in my view, this is even more problematic because this seems to be happening for good reason.

We are currently confronted with the apparent incapacity of modern liberal democracies to meaningfully prevent the effects of climate change. Leaving the problem of collective action to one side, there is another widespread explanation for this failure to act: short-term electoral pressures tend to dictate the agenda of a democratically elected government at the expense of longer-term concerns – such as climate change (see Compston 2009, Compston and Bailey 2008). This is because longer-term decisions tend to involve short-term costs that most citizens are (at least, presumed to be) unwilling to pay, since they do not receive any immediately related gains. On the most pessimistic versions of this view, the short-sightedness of citizens in democracies structurally binds their leaders into making poor and, ultimately, unsustainable political decisions in order to stand a reasonable chance for re-election.

As a result, the thought occurs that we might be better served were we ruled by experts rather than by representatives of the people. After all, were we to entrust the reins of power to the experts, to those who know, to the scientists and learned people who truly understand the causes and consequences of climate change, we could plausibly expect to face far improved prospects for effective action being taken to meet this very real threat, principally because such experts would have access to all the relevant knowledge to address the problem without ever having to stand for re-election. The moral imperative seems convincing, since such a change might save many lives and prevent much human suffering. It is in this context that the temptation of epistemic elitism is strong. And Plato elegantly expresses the root of this appeal in *The Republic.*

Plato (1987: Books V and VI) begins by establishing a distinction between philosophy and opinion by relying upon a presumed implacable hierarchy of knowledge, where more abstract, more rigorous, more logical beliefs are to be preferred to the more intuitive and common beliefs of the *hoi polloi*. Plato offers a simple but powerful reason in support of this classification: the former offer truth and are based on good reasons, whereas the latter offer potential falsehoods and are mostly based on prejudice. Thus, according to Plato, it is our commitment to truth which requires that we turn our attention away from the muddled thoughts and beliefs held – and all too often cherished – by mere lay people. Plato thus goes on to make the argument for an enlightened form of authoritarianism (or as we might call it today, epistemic elitism) spearheaded by knowledgeable and well-trained philosopher-kings on the basis of this conception of knowledge (1987: Book VI). The general argument he offers in support can be summarised thus: since some members of society have privileged access to reliable knowledge (call them members of the epistemic elite), it makes sense to entrust the responsibility to guide and direct the affairs of the state to them, because they will be able to make best use of such knowledge to arrive at informed and reasoned decisions when ruling society. Hence, the attraction to this form of government in our present context stems from our apparent need for strong rulers, willing and capable of acting on the basis of the scientific facts irrespective of public opinion.

In response, committed democrats must rise to the challenge. It will not suffice to appeal to the inherent worth of democracy or to its historic benefits. We must address our real and present problems in order to meaningfully justify keeping the democratic experiment alive. And for this, we must provide more than vapid calls for a change of mindset, conduct or culture. We must provide a vision of how such changes could reasonably be expected to come about.[3] Philosophers are in the business of providing such visions and that is why I think philosophy can help. More specifically though, I think John Dewey's political philosophy contains a democratic ideal that can be made relevant to our present problems. That is why I will now turn to a discussion of the main tenets of this ideal.

Deweyan Democracy as a Guide to Devising Cultural Responses to Collective Problems

> "In its most vibrant sense, democracy provides an ideal of community life in which citizens engage in social discourse to determine how projected actions will impact on others" (Boisvert 1998: 57).

The term 'democracy' generally refers to a mode of government which relies, in some way or another, on popular involvement in political decision-making. In its ancient form, democratic participation was direct, with the public as a whole actively involved in collective deliberations. Yet, over time, growth in the size

---

3   Of course another tack philosophers could take would be to criticise the authoritarian temptation by pointing to the potential risks associated with it. But I think this strategy will not suffice because even the worst authoritarian transgressions are likely to end up paling in comparison with the threats associated with inaction in relation to climate change.

of the public has made it impossible to envision such direct forms of popular participation effectively shaping major political decisions. With a citizenry of several millions, with complex bureaucratic institutions governing an extended territory and with heavy reliance on highly sophisticated technological expertise, the nature of modern democracy seems irreconcilably at odds with the participative democracy of the ancients. And this because citizens of advanced democracies tend to live too far away from each other, tend to lack the necessary leisure time for engaging in public affairs, and they all too often fail to understand their system of government and the technologies at their disposal, to competently and meaningfully engage in the task of ruling (Wolff 1996: 68–103).

Contemporary liberal democracies hence rely on public representatives to do this instead. The participation of the *demos* is mostly limited to voting in regular elections for their representatives. This system is presumed to ally the virtues of epistemic elitism (in the sense that rulers are likely to be highly educated and have access to the time and knowledge necessary to make intelligent decisions) with the virtues of popular rule (since the people can, at election time, rid themselves of leaders they deem to be unsatisfactory) (Wolff 1996: 103–112).

But in our present context we can see that liberal democracy still seems to suffer from a revised version of the worry raised by Plato: Are the representatives of the people mandated to follow public opinion and therefore seek to channel the wills and desires of the masses? Or, are they there to exercise leadership, making when necessary potentially unpopular decisions?

It was in response to this problem, and in particular to the elitist solution offered by Walter Lippman, that Dewey wrote *The Public and Its Problems* in 1927. Lippman had forcefully argued that the complexity of modern democracies requires that the mostly uneducated public be kept away from the essentials of government, leaving to competent and neutral experts the task of setting day-to-day policy. Why? Mostly because the public's ignorance and limited availability of time forbid it from making relevant and useful contributions to the task of policy-making. Lippman (1925) therefore argues that citizens ought to leave the domain of political deliberation to the experts.

Although Dewey accepted the diagnostic, he responded by fundamentally rejecting the purported cure. What we need in order to improve our democracy, Dewey argued (LW2: 262–494), is not so much the benevolent intercession of experts, but the conditions under which an educated and engaged citizenry can come to life. In his view, the key democratic resource is the collective intelligence of an active, problem solving citizenry.

But here already we must pause to clarify: Dewey did not advocate a return to the direct form of democracy of the ancients; he fully recognised the need for public representation in the contemporary context. However, the object of his fundamental insights was not so much the apparatus of government as it was the broader realm of human association (LW2: 325, LW11: 25). In fact, his work predominantly deals with the role citizens could and should play within the civil society of a liberal democracy. This is because Dewey thought democratic deliberation ought to happen throughout society, not just in the deliberating assemblies of the state. We should therefore understand him as attempting to make sense of the place and promise of popular participation *within* the context of representative democracy.

More specifically though, Dewey sought to address the nodal points where smaller forums for deliberation exist. He identified "the family, the school, industry, religion" (LW2: 325) along with the rest of civil society as crucial sites where true democracy could and should be fostered. But, evidently, such a view presupposes a particular notion of what 'true' democracy consists in which I have yet to explain. So, I must now turn to a more substantive discussion of Dewey's democratic ideal.

Active Citizenship or Democracy as a Way of Life

> "Participation in public life doesn't mean that you all have to run for public office [...] But it does mean that you should pay attention and contribute in any way that you can. Stay informed. Write letters, or make phone calls on behalf of an issue you care about. If electoral politics isn't your thing, [...] find a way to serve your community and your country – an act that will help you stay connected to your fellow citizens and improve the lives of those around you" (Obama, 2010: 'Michigan Graduation Speech').

According to Dewey, the more common understanding of democracy (bequeathed unto us by classical liberalism) as a primarily institutional arrangement has for principal aim to defend the selfish rights of individuals and it thus fails to make sense of the need to actively encourage democratic participation. Crucially, he contends that classical liberalism is based on the erroneous assumption that human nature is set independently of social institutions, and that the role of the democratic state is to protect the pre-given interests of atomistic, selfish individuals (LW5: 42–125). In contrast, Dewey tells us that individuals are not pre-given; the institutions and communities in which they are raised fundamentally shape them. Or as Dewey puts it, "to learn to be human, is to develop through the give-and-take of communication an effective sense of being an individually distinctive member of a community" (LW2, p. 332). It is thus on the basis of this conception of the person as socially-encumbered, as inescapably dependent on others, as embedded in and produced by the relationships entertained with others, that Dewey develops his conception of the democratic ideal. He thus writes:

> "From the standpoint of the individual, [democracy] consists in having a responsible share according to capacity in forming and directing activities of the groups to which one belongs and in participating according to need in the values which the group sustain. From the standpoint of the group, it demands liberation of the potentialities of members of a group in harmony with the interests and good which are common" (LW2: 327–328).

We can glean from this passage that, for Dewey, democracy is primarily a form of collective deliberation, where people engage each other in conversation within and across existing groups (say, workers' unions, school meetings or a religious action groups) to come to communal decisions. This is a highly devolved notion of democracy, where the key feature of democratic life is to be found in the social and moral fabric of citizens, not primarily in the institutions under which they live. Therefore, on this conception, democracy is a "social" and "moral ideal" in which public deliberation is not just permitted, but encouraged in all spheres of human association (LW2: 325, LW7: 349).

But left at this, it seems that Dewey does not really offer a reply to the elitist challenge. Although he tells us that citizens should engage in the democratic process throughout society, it is not clear how their efforts ever come to weigh upon actual policy-making. Without this, it appears that the democratic citizens Dewey has in mind would be active, but ineffective. Rather like an impotent wasp endlessly buzzing in the face of an aggressor, but never managing to sting. So, we must ask: how are Deweyan citizens supposed to sting?

Democracy as Applied Social Intelligence

"The method of democracy – inasfar as it is that of organized intelligence – is to bring these conflicts out into the open where their special claims can be discussed and judged in the light of more inclusive interests than are represented by either of them separately" (LW11: 56).

On the Deweyan story, deliberation in the varied spheres of associative life is hoped to come to bear on actual policy-making by generating a vibrant public. This would be done by: (a) educating citizens about contemporary political issues; (b) generating actual improvements of the piecemeal decisions arrived at as a result of deliberations; (c) organising citizens into coherent groups to get their collective voices heard on particular issues; and (d) fostering a spirit of shared purpose between all citizens (Boisvert 1998: 73–94). Taken altogether, this amounts to transforming the citizenry into a citizens' army of resourceful activists with both the ability and the desire to: (i) convince and lobby their elected representatives into consulting and listening to them on specific problems and matters of general policy; and (ii) bring about change through social action within the sphere of civil society. The experimental and somewhat haphazard nature of this vision of democracy is considered to be its core virtue. For, according to Dewey, the scientific knowledge of experts remains incomplete until it is placed "in the context of a future which cannot be known but only speculated about and resolved upon" (MW11: 48). The many deliberations of citizens are thought to contribute in the task of putting scientific knowledge into such a context.

Thus, at the heart of this vision, we find the idea that the manifold forms of involvement in joint ventures seeking to address the great variety of practical problems confronted throughout society would have a cumulative effect in shaping public debate. But, for Dewey, this is only the case if such diverse groups of citizens are united in embracing a common standpoint to address their many problems, namely: the standpoint of applied social intelligence.

Dewey claims (LW12: 11) that all deliberation begins with the external disruption of business-as-usual (i.e. our pre-reflective, habitual behaviour). An obstacle of some kind or another must emerge, stopping us from continuing to act according to our habits of action. We are thus confronted by the need to revise our behaviour to continue to get what we want. Dewey advises that we envision that the resolution of such problems can take two forms: (1) the overcoming of the obstacle to return to our past behaviour, or (2) the abandonment of the pursuit of the goals that can no longer be achieved because of this obstacle. Thus, if we were to take option (1), this is how we could reconstruct the process of inquiry:

If a set of facts F comes about which does not allow us to achieve our ends E; then, we must inquire into the nature of the situation, in order to decide which action A, will result in the factual situation F', which enables us to achieve E.

> However, this highly schematic account of inquiry fails to represent what Dewey considers to be a crucial part of this process. Indeed, for him option (1) and (2) are not completely impermeable (LW12: 108–118).[4] In the face of many problems, we compromise by revising our ends, in such a manner as to be able to achieve our new ends-in-view in this new context. Thus, a dialectical relationship exists between our two options, giving our scheme a slightly different look:

If a set of facts F arises in which we can no longer achieve our ends E by relying on our habits of action; then we must therefore inquire into the nature of the situation, and then decide whether to continue pursuing E, and perform action A, or opt for E' and perform action A'.

According to Dewey's democratic ideal, members of the public must be involved in precisely this task of discussing the possible revision of both our ends and the means to pursue them, principally because it makes for much improved decisions. And this is where the superiority of democracy in relation to epistemic elitism comes to the fore: experts on their own can only ever offer advice about the *means* to fulfil fixed ends, for their expertise is only limited to questions of means; but to respond in the most intelligent ways to our problems the very *ends* that we pursue must also be up for discussion. For this flexibility of ends and means affords us with a richer array of potential solutions to consider. How so? By enabling all

---

4   Specifically, Dewey writes (LW12: 111): "Organic interaction becomes inquiry when existential consequences are anticipated; when environing conditions are examined with reference to their potentialities; and when responsive activities are selected and ordered with reference to actualization of some of the potentialities, rather than others, in a final existential situation. Resolution of the indeterminate situation is active and operational. If the inquiry is adequately directed, the final issue is the unified situation that has been mentioned."

those who are affected by a problem to express their concerns, and to take part in the discussions seeking to address the problem, the decision-making process can draw on their specific experiences, ideas, and aspirations to broaden the field of possible alternatives in responsible and responsive ways. In other words, this kind of participation allows us to hope that the interaction between participants can generate more creative and adapted solutions, through the collective revision of both our ends and the means we use to achieve them in context-sensitive and creative ways (see Festenstein 2001).

Climate change clearly stands to be a major interruption to business-as-usual, that is why applying this model to the great variety of specific problems raised by the threat of climate change offers a dynamic vision of how the institutional changes made by governments can potentially come into dialogue with the everyday (following the definition developed previously, one might say 'cultural') concerns and habits of citizens. In other words, I think Deweyan democracy offers a promising cultural response to climate change because:

- climate change presents us with a great many complex problems that have the potential to adversely affect all of humanity; and,
- Dewey's deliberative politics offers a model for cultural and institutional change which could provide effective and responsive solutions to the problems caused by climate change.

However, seeking to apply Dewey's democratic ideal is no simple affair. In order for citizens to be well equipped when engaging in the task of applied social intelligence, citizens must be able to bring their collective needs and aspirations into dialogue with the concrete limitations they face in any given context. To provide citizens with the necessary capabilities to do this, Dewey looks to education.

"It's the education, stupid!"

> "The citizen should be moulded to suit the form of government under which he lives. For each government has a peculiar character which originally formed and which continues to preserve it. The character of democracy creates democracy, the character of oligarchy creates oligarchy; and always the better the character, the better the government" (Aristotle 1981: 1337a11–17).

The title of this sub-section of course refers to Bill Clinton's famous 1992 campaign strategy to focus on the economy. Although such a rhetorical move proved to be *electorally* successful, the key Deweyan insight is that good democratic *government* requires a meaningfully educated citizenry. More specifically, Boisvert (1998: 93) writes that the Deweyan notion of intelligence in action involves the following components: "1. Awareness of fundamental principles"; "2. Attention to consequences of varied actions"; "3. Information about contemporary issues";

"4. Dissemination of the results of inquiry"; "5. Transforming all of this into instrumentalities for social change."

Thus, in order, to properly and fully participate in the task of collective problem solving, citizens must be morally sensitive, practically minded, meticulous, well informed, pro-active individuals with access to effective broadcasting means to air their views. Clearly, much needs to be said to clarify what each of these properties amount to in practice. But for the sake of this chapter, it will suffice to say that, on Dewey's account of democracy, citizens need to have a broad set of intellectual and moral dispositions enabling them to intelligently and responsibly engage with the full complexity of their collective problems. Evidently, the overwhelming majority of citizens in modern democracies fall short of this description. This is problematic insofar as it suggests that Dewey's democratic ideal might be of no immediate use to us after all.

However, Dewey crucially tells us that democratic changes can only be expected to be as good as the citizens affected by them are caring, informed and active. The key democratic task then is to encourage the development of an intellectually, morally and practically engaged citizenry. And the key social institution Dewey identifies in this task is the school (EW5: 93, MW10: 139).[5]

Typically, when envisioning the role of the school, two models of education with two very different ends are traditionally offered to us. On the one hand, we are presented with a conception of education where knowledge is valued for its own sake and learning essentially consists in the amassment of knowledge. On the other hand, we are presented with the notion that education is of value only inasmuch as it prepares children to become productive members of society and learning consists in the acquisition of skills which will be of value to individuals in the market place. Both of these models prepare students for two forms of disengaged citizenship: the first, by creating a false disconnection between knowledge and culture (or as we often say in philosophy, between theory and practice), resulting in the failure to put knowledge to social use; and the second, by fostering a selfish notion of practical engagement with the world, single-mindedly focused on the prospects of pecuniary gain, resulting in the inability to meaningfully cooperate with others.

Dewey rejects both these conceptions of education and argues instead for a form of schooling designed to develop active citizens by cultivating in them the dispositions and habits of communal problem solving. Thus, in addition to the acquisition of basic skills and knowledge, students in a Deweyan school would be encouraged to learn to think, dialogue and work towards solving communal

---

5   Specifically, Dewey writes: "I believe that education is the fundamental method of social progress and reform" (EW5: 93); and: "In a complex society, ability to understand and sympathize with the operations and lot of others is a condition of common purpose which only education can procure. The external differences of pursuit and experience are so very great in our complicated industrial civilization, that men will not see across and through the walls which separate them, unless they have been trained to do so" (MW10: 139).

problems through collective deliberation. By growing accustomed to this process, students would acquire the dispositions and habits that would later enable them to become active problem solving citizens in wider society (Dewey, MW9: 1–375).

Mathew Lipman developed this notion further still and spoke of the classroom as a "community of inquiry" (2003: 20). In this community of inquiry, the teacher is not primarily tasked with the transfer of knowledge to pupils; she is rather engaged in the orchestration of discussion among them. By exposing students to intellectually stimulating material, by supporting them when expressing their views, by encouraging others to meaningfully and respectfully respond, the teacher is instilling in her pupils the dispositions to thoughtfully engage with important problems and to take part in genuine communication, by entering the process of exchanging reasons with one another. And according to the educational program developed by Lipman (called 'Philosophy for Children'), if this is done well, after a time, students themselves will be able to orchestrate this dialogue with only minimal intervention from the teacher. Thanks to this practice, students are expected to develop the ability to think:

- Critically, by thinking thoroughly about the issues at hand (M. Lipman, 2003: 205–242);
- Collaboratively, by engaging in genuine collective discussions, learning how to listen to others and to make one's self heard (2003: 83–125);
- Caringly, by addressing morally important issues, but also by engaging in a respectful and open dialogue (2003: 261–271);
- Creatively, by enabling the flow of ideas and arguments to take a relatively free course, with no pre-given endpoint – simply a commitment to the process of discussion itself (2003: 243–260).

It seems plainly evident to me that such are the dispositions of thought active citizens will need to adopt to meaningfully engage in piece meal efforts to prevent and/or mitigate the effects of climate change. Why? Mostly because the scope, complexity and the relative unpredictability of the problems we are likely to encounter as a result of climate change generate the need for a citizenry made up of individuals able and willing to make well adapted and morally sensitive judgments. In the face of climate change, the need for an active and educated citizenry is no longer based on the leisurely hope of improving human life; it is based on the most serious need to preserve it.

And yet, most national curricula in modern democracies continue to fail to put collective problem solving at the centre of our children's education. Consequently, I think we need to focus some of our efforts on reforming our schools in order to prepare our children to eventually become responsible citizens. In my view, effective cultural responses to climate change would be more likely to sprout up and grow to their full potential in a society made up of citizens nurtured in the spirit of Deweyan democracy.

Conclusion

In summary, in this chapter, I have sought to show how the Deweyan democratic ideal could provide a general vision of how cultural responses to the many problems related to climate change might be meaningfully encouraged. To do this, I started by suggesting that philosophy could play a role in the task of imagining ways in which cultural change can be brought about. In particular, I pointed out that although philosophy could suggest a rather authoritarian approach to solving the problems we currently face, it could also provide a vision of how we might be able to confront our problems within a democratic context via cultural reform. Most crucially, I identified John Dewey's conception of democracy as the best candidate to do so. I then offered a more detailed account of Dewey's conception of democracy. And I thus argued that Dewey's democratic ideal provides an appealing understanding of the process in which active and educated citizens could come together to fruitfully contribute in the task of piecemeal cultural reform in the face of common problems, such as those generated by climate change. Furthermore, I claimed that particular emphasis should be placed on the need to nurture active citizens into existence through particular educational practices (such as, 'Philosophy for Children'). Finally, I suggested that this could potentially help us in the task of providing cultural responses to climate change primarily because it would prepare our citizenries to meaningfully engage with the full magnitude and complexity of the problems associated with this crisis.

As a closing note, I must admit that although the aim of this chapter is to offer a hopeful prospect for cultural improvement in the face of climate change, I fear that the time for melioristic optimism may well be rapidly running out. My original intention was to highlight the importance of taking an educational or developmental approach to collective problem solving in the face of climate change. However, there is a fundamental tension pulling against the thrust of my argument: education is a relatively slow process, while climate change is an urging crisis. With every day driving us closer to major climatic tipping points, the case for an educational approach to the problem weakens. As the crisis intensifies, active and informed citizens are likely to come to the obvious conclusion that we need more than mere faith in education or in collective deliberation: we need concrete solutions to pressing human problems.

Bibliography

Aristotle. 1981. *Politics*. Revised Edition. London: Penguin Classics.

Bates, B.C. Kudzewicz, Z.W. Wu, S. and Palutikof, J.P. (eds). 2008. *Climate Change and Water: Technical Paper on the Intergovernmental Panel on Climate Change*. Geneva: IPCC Secretariat.

Boisvert, R. D. 1998. *John Dewey: Rethinking Our Time*, Albany: SUNY Press.

Brown, L.R. 2008. *Plan B 3.0: Mobilizing to Save Civilization*, London: Earth Policy Institute.

Brown, L.R. 2009. Could Food Shortages Bring Down Civilization? *Scientific American Magazine* [Online]. Available at: http://www.sciam.com/article. cfm?id=civilization-food-shortages [Accessed: 26/04/2009]

Cicerone, R. J. 2006. Geoengineering: Encouraging Research and Overseeing Implementation, *Climatic Change*, 77 (3–4), 221–226

Compston, H. and Bailey, I. (eds) 2008. *Turning Down the Heat: The Politics of Climate Policy in Affluent Democracies*, Basingstoke: Palgrave Macmillan.

Compston, H. 2009. The politics of climate policy: strategic options for national governments. Paper prepared for *5th ECPR General Conference*, Potsdam, 10–12 September. Available at: http://www.cardiff.ac.uk/euros/resources/ The%20politics%20of%20climate%20policy.pdf [Accessed: 17/03/2010].

Crutzen, P. J. 2006. Albedo Enhancement by Stratospheric Sulfur Injections: A Contribution to Resolve a Policy Dilemma?, *Climatic Change*, 77 (3–4), 211–220.

Dewey, J. 1969–1990. *The Collected Works of John Dewey: The Early Works,*

The Middle Works, The Later Works. 37 volumes, edited by JoAnn Boydston, Carbondale (Illinois): Southern Illinois University Press – abbreviated as respectively EW, MW and LW.

Festenstein, M. 2001. Inquiry as critique: On the legacy of Deweyan pragmatism for political theory, *Political Studies*, 49 (4), 730–748.

Homer-Dixon, T. 1999. *Environment, Scarcity and Violence.* Princeton (New Jersey): Princeton University Press.

Homer-Dixon, T. 2007. Positive Feedbacks, Dynamic Ice-Sheets, and the Recarbonization of the Global Fuel Supply: the new Sense of Urgency about Global Warming, Chapter 2 in *A Globally Integrated Climate Policy for Canada,* edited by S. Bernstein et al. Toronto: University of Toronto Press.

International Panel on Climate Change. 2007. *Climate Change 2007 – Impacts, Adaptation and Vulnerability*, Contribution of Working Group II to the Fourth Assessment Report of the IPCC. Geneva: IPCC Secretariat.

International Panel on Climate Change. 2007. *Climate Change 2007 – The Physical Science Basis,* Contribution of Working Group I to the Fourth Assessment Report of the IPCC. Geneva: IPCC Secretariat.

Kraut, R. 2009.Plato, *The Stanford Encyclopaedia of Philosophy.* Available at: http://plato.stanford.edu/entries/plato/ [accessed: 15/03/2010]

Lippman, W. 1925. *The Phantom Public.* New York: Harcourt.

Lipman, M. 2003. *Thinking in Education.* Cambridge: Cambridge University Press.

Monbiot, G. 2009. Climate change displacement has begun – but hardly anyone has noticed. *The Guardian* [Online] 08/05/2009. Available at: http://www. guardian.co.uk/environment/georgemonbiot/2009/may/07/monbiot-climate-change-evacuation [Accessed: 16/03/10]

Obama, B. 2010. *Michigan Graduation Speech*. Michigan: 1st of May 2010. Available at: http://www.huffingtonpost.com/2010/05/01/obama-michigan-graduation_n_559688.html [Accessed: 02/05/2010]

Plato. 1987. *The Republic*. Revised Edition. London: Penguin Classics.

Plato. 1993. *The Apology* in *The Last Days of Socrates: Euthyphro; The Apology; Crito; Phaedo,* edited by H. Tarrant, London: Penguin Classics, 29–68.

Stern, N. 2007. *The Economics of Climate Change*. Cambridge: Cambridge University Press.

Wolff, J. 1996. *An Introduction to Political Philosophy*. Oxford: Oxford University Press.

Sustainability Means Ethics and This is a Cultural Revolution

Michel Puech

Something is fascinating in the very idea of climate change as leading to a cultural change. Our visions of the future are disrupted. Our views on science and on progress are transformed by an unexpected newcomer in this civilization of power and domination: the fear of a self-provoked collapse. But a growing suspicion insinuates that the fear is perhaps a little bit untimely. Is a deep change really necessary? How deep? My thesis is about the nature of the change that sustainability calls for. I argue that this change is *ethical* and not political, economical, or institutional (Puech 2010).

Changing ... What?

Hitherto, changing and evolving was a natural property of life, a Darwinian process. Species evolved through a continuing pressure of competition to survive. Surviving is the basic and biological form of sustainability. With the Homo Sapiens, a new engine of evolution and a real booster of change began: culture and its two change facilitators of unknown might, language and technology. Since we talk and think, since we build artefacts and rely more and more on them, our evolution is more and more cultural, less and less natural. This supremacy of cultural evolution over natural evolution does not follow from superiority in essence as philosophers would say. It is just a question of timing. Techno-evolution runs incredibly faster than natural evolution, the latter being the evolution of species and the evolution of the ecosphere and its balances. Here, we tumble upon ecological sustainability issues. Most of them are troubles caused by the accelerated pace of change that the human species imposes. One of them is the climate change concern. Climate science tries to make people understand that the problem is not the fact that the climate is changing because of us, but that it is changing so fast that the instability is unpredictable and potentially dangerous (I hope that this moderate statement will not be construed as climate-scepticism, even if I adopt a moderate epistemological scepticism on every subject, including this one).

Human responsibility for a massive ecological change is not a recent scientific finding. It is not discovered though sophisticated computer simulations. It was accessible long before the 1972 'Club of Rome' report, before the revered 'whistleblowers' of the 1960s (R. Carson, L. Whyte, B. Commoner, P.R. Ehrlich, G. Hardin), and even before W. Vernadsky's founding of global ecology (Vernadsky 1926). A gentleman born in 1801, George Perkins Marsh, published a book in 1864 on the effect of man on nature. Its 'humble pages' do not aspire to qualify as science, as the author specifies, but Marsh's conclusion from experience and simple observations is unambiguous: 'But we are, even now, breaking up the floor and wainscoting and doors and window frames of our dwelling, for fuel to warm our bodies and seethe our pottage, and the world cannot afford to wait till the slow and pure progress of exact science has taught it a better economy.' (Marsh 1864: 52). For some in the nineteenth century and before, and now for everyone on this planet, the issue is that our power to modify the natural world is essentially destructive. No technoscientific magical solution will arise, and we know that the pace of this change will lead to an ecological collapse.

We are not in the cultural phase of discovering the need to change our industrial behaviour. We are in the very different phase of facing the consequences of unmade changes. We are facing a 'revenge' of Nature against a permanent aggression. We were aware of this aggression, but we decided not to consider it because we were perfectly convinced that we were the strongest. When the climate crisis is construed as a 'Revenge of Gaia' (Lovelock 2006), the need for a change is not only seen as an opportunity. It is experienced as a perhaps undeserved last chance.

The sustainability challenge is to match cultural and natural change. On the one hand, science tries to understand, to model, and predict natural change. On the other hand, the humanities and the relevant academic community have to think and elaborate on cultural change, but now in the new light of the sustainability change. The triggering factor is the awareness that our technoscientific culture is not sustainable. That is, it leads to its own extinction, for ecological and/or economic and social reasons.

We are challenged, but is it by the climate? Not only, for sure, and not essentially, I believe. Is it by the Earth, an abandoned Deity whose 'revenge' is frightening? What we mean by this unnecessary deification of the planet is that the biological and symbolical dimensions of our culture and representation of values have been upset. But one step further in awareness is required. It is an ethical issue that we face, an assessment of our modern self in its relation to itself, to its integrity. Therefore, philosophy and not only a functional approach is needed to respond to the intellectual challenge of sustainability It is fundamentally a cultural change, which is not only a matter of official 'sustainability policy'. It requires a sustainability ethics. It implies not only a political but also a cultural revolution. This hypothesis offers an explanation for what remains so difficult to understand and to accept: why sustainability politics are ill-fated and why institutional manoeuvres necessarily produce blame-avoiding policy and nothing else (except taxes, as sure as death in the end). I do not intend to demonstrate this failure *a priori*. I merely aim to take into account an a posteriori matter of fact. Since the 1950s we have created dozens of international institutions devised for handling political, economical, social and ecological global issues. They have led to unquestionable achievements and the question is not on assessing their efficiency as a whole. My point is to focus on this a posteriori and factual observation: *current sustainability issues are those that resist current institutional treatments*. In this sense precisely, we are facing a new kind of civilization problems. A couple of years ago I felt isolated when I said so, but since the Copenhagen failure (Cop 15) and the Grenelle failure (an ambitious stake-holders consensus program that boiled down to nothing) in my country, France, I sense I am making new friends.

This chapter supports the ethical priority in sustainability with two types of argument: a negative and extrinsic argument about institutional failure, and a positive and intrinsic argument about the nature of sustainability as self-reliance.

The Invisible Collapse ... of Institutions

I start with a statement of facts. We are currently attempting a sustainability reform and it does not seem to work. When existing power structures plan and enact change, it is a reform. Otherwise, we are on our way toward a revolution, a change *of* governing structures and not a change *by* governing structures. I will suggest later in this chapter that this is a new kind of revolutionary change. It is micro-political to the extreme. It is ethical. But for now, let us look again at the facts: sustainability change as institutional reform does not work. The logic of 'small steps in the right direction' is now totally worn out, in my opinion. There is no better case study for this argument than Cop 15. For years, in every administration concerned with sustainability and in political studies departments, a constant flow of elaboration and bureaucratic literature on Cop 15 drenched the actors of sustainability. The media advertised the event or the politicians' participation in it. The day after was a real 'day after', a sudden downsizing of expectations and communication, reduced to almost nothing except the usual frail and unconvincing claim of 'small steps in the right direction'. In the so-called governance of sustainability issues, we have invented a new paradox of change and movement: small steps that do not drive us any closer to the target – I mean a minimal and consensual target, accepted by governments, international institutions and NGOs in their admirable texts and declarations. The same assessment applies to the 'symbolic' change argument. It looks as if we have found a way to use symbolic in order to replace instead of promoting real change – I mean pragmatic and factual change in material consumption or production processes, in transportation, disposal and waste management and other humble activities with strong ecological impact. Discourse and communication have changed. Anything can change as long as it remains symbolic and iconic action.

I take 'institution' as the name for a collective entity whose power and interest systematically predominates over those of its individual human members. Nation-states are institutions par excellence, but also the UN, any government department or agency, a local community lead by a professional politician, large firms a NGO and so forth. Almost everyone applies the *institutional paradigm* with no idea of any alternative. It says: 'solutions come from institutions'. In face of any concern (health, education, moral dilemma, etc.), the question 'how to cope' is spontaneously translated into 'what is the institution to delegate to?' Some sub-concerns follow from this approach: how to improve this institution, its efficiency, how to lobby, to suggest rules and regulations, and so forth.

Instead of the possible ecological or economic collapse, our priority should be the actual institutional collapse. It deprives us of the means to cope with any other menacing crisis. Institutions will not walk the talk and they never intended to. They channel militant energy to move the cogs that move other institutional cogs, and everything is in order as long as the energy remains inside the institutional machine and does not threaten to damage its functioning. If we had a functional problem, the solution would be a functional solution, that is to say another content for our institutional machines. But the problem is a meta-problem about our functional approach itself. Then every institutional machine is part of the problem. This may explain the uncomfortable impression we have of tossing and turning with an inexhaustible energy while remaining locked inside the problem.

Instead of Jared Diamond's version of the collapse (Diamond 2005), I will embrace J.A. Tainter's analysis of 'the collapse of complex societies' (Tainter 1988). From a significant list of civilization collapses through history, Tainter concludes:

> The collapses of these societies cannot be understood solely by reference to their environments and subsistence practices (or to changes in these), to the pressure of outside peoples, to internal conflict, to population growth, to catastrophes, or to sociopolitical dysfunction. What affected the Romans, Mayans, and Chacoans so adversely was how one or more of these factors was related to the cost/benefit ratio of investment in complexity. When challenges and stresses caused this ratio to deteriorate excessively, or coincided with a declining marginal return, collapse became increasingly likely. (Tainter 1988: 187).

Declining marginal returns of institutional sophistication is exactly what we are experiencing in sustainability. We invest in ever more institutional sophistication to achieve ever less.

This pessimistic approach is not necessarily an incentive to radicalism, marginality or violent action. I take it as an incentive to *really renounce technocracy*. This implies a bottom-up cultural revolution, which is not an institutional revolution because it happens on a different level. It does not challenge the existing institutions but boldly ignores them. To care for nature is not to care for a bureaucratic process that cares for nature. To care for climate is not to care for the UN institutions in charge of climate. Nordhaus and Shellenberg used the provoking subtitle: 'Why We Can't Leave Saving the Planet to Environmentalists' (Nordhaus, Shellenberg 2007). We must also understand: Why We Can't Leave Saving the Planet to Institutions.

But in this case, *To whom* can we leave saving the planet? That is the question. The answer from sustainability ethics is: to no one, it depends on you. Abandon the idea of 'leaving it to someone', the idea of *delegation*. Delegation politics has proved its limits in the field of sustainability more than in any other. I believe that a re-*appropriation* of sustainability as an ethical concern is at the centre of the change for sustainability.

Cultivating Satiety and Self-Reliance

This revolution is not a brutal shift of power. It is a slow and enduring bottom-up change. It is essentially an empowerment of micro-actors. Much has still to be invented, but excellent tools are available for sustainability conceived as ethics.

In *Walden* (Thoreau 1854), Henry David Thoreau provides a pattern of self-reliance ethics that is at the same time political economy. Thoreau's entire philosophy is a model for change towards sustainability, an ethical change, once every hope of institutional change has been lost: 'The true reform can be undertaken any morning before unbarring our doors. It calls no convention. I can do two thirds the reform of the world myself.' (Thoreau's *Journal* quoted by R.D. Richardson 1986: 106). The triple formula of Thoreau's ethics is far simpler than Kant's. It is: 'Simplicity, simplicity, simplicity!' (Thoreau 1854, chapter 'Where I Lived').

Inspired by Thoreau's ethics of self-reform, Gandhi achieved a major political change in his own country. In Gandhi, we find the paradigm of an ethical reform that causes by its own impetus a major institutional change. I believe that what the official sustainability politics intend to do is exactly the opposite: an institutional change that may bring about, by conviction or by obligation, a quasi-ethical change, a change in the *ethos*, the individual principles and ways of life. Gandhi's way is the opposite. Its roots are in the *satyagraha* attitude, the personal aspiration to truth and, more than that, the personal striving for authenticity (*satyagraha* can be defined as a self-reliant and uncompromising passive resistance). Here again, self-reliance and frugality are intimately tied. Gandhi's *swadeshi* movement was a cultural change in economy and political economy. It aimed at the material independence of a community, the material sustainability of a local community, as far as possible *(swadeshi* can be defined as the realization of a global economic strategy through personal actions of production/consumption).

The volume 5 of Arne Naess's *Selected Works* (founder of 'deep ecology' and a professional philosopher) bears the title 'Gandhi and Group Conflict: Explorations of Nonviolent Resistance, Satyagraha' (Naess 2005). Naess understood that the global challenge of our culture was to achieve a revolution through ethical self-reform. This is the meaning of 'deep' in a philosophically acceptable *deep ecology*. It still conveys radicalism and revolution, but not as a politically aggressive movement. The cultural form of global change is to be ethical, based on consistent self-governance and personal awareness. Historic propagandists confirm this view: 'Some of the most far-reaching changes are coming from the grass roots as individuals see their lives and their relationships with nature in a new light. As a result, they are making changes in their life-styles and are insisting on changes in public policy.' (Brown et al. 1991: 166).

Satiety and self-reliance are not heroic virtues, said the founder of the *voluntary simplicity* movement, R.B. Gregg, another disciple of Gandhi: 'Our present "mental climate" is not favourable either to a clear understanding of the value of simplicity or to its practice. Simplicity seems to be a foible of saints and occasional geniuses, but not something for the rest of us.' (Gregg 1936). The ethics of sustainability is an ethic of ordinary life. It lies in micro-actions of care and awareness. These values have nothing to do with a pathological need to be famous as a virtue champion and exceptional ascetic performer. Rather, and this is the deep side of ethics, sustainability is grounded on the very sane need *to be*, just to be: not to survive through sacrifices, but to be a human person that takes responsibility for him/herself. A person is built in a constant effort of self-responsibility. This conception of the sustainable self is a quest for every person, all life long, across experiences and achievements, findings and disappointments, the multitude of micro-events captured in ethical awareness. The 'voluntary simplicity' movement and D. Elgin in particular, confirm this ethical and metaphysical substance: 'To live more *voluntary* is to live more deliberately, intentionally, and purposefully – in short, it is to live more consciously.' (Elgin 1993: 24). Instead of global values to be revered, sustainability ethics promotes intimate values to be discovered. 'The particular expression of *simplicity* is a personal matter. We each know where our lives are unnecessarily complicated.' (*ibid.*). Instead of Hans Jonas's ethical oligarchy (the power given to a 'responsible' elite in order to 'save' the rest of us), which is nothing else than domination-as-usual, the German philosopher Dieter Birnbacher suggests a brilliant hypothesis on the nature of a modest personal ethic of the future: responsibility for the future is not a new and not even a particular ethic. It just stresses the very nature of ethics (Birnbacher 1994: 87).

Thus, the sustainability cultural revolution is a rotation movement, according to the original meaning of 'revolution'. It drives us back to ethics in itself, to the simple idea of an ethical dimension in our personal life, to the consequences on community behaviour of this change, and, last but not least, to its consequences on the human footprint on this planet.

Sustainability: The Ethical Turn

According to this micro-political and anti-institutional approach, sustainability is not *a* but *the* ethical turn in our global culture. It is an ethical turn because what has to change is one's behaviour. What is required from us is a change of *ethos*. But there is more. In the end what was unsustainable in our modern cult of growth and power was nothing but the loss of ethics in our collective and personal ethos. What is essential in the cultural change induced by sustainability is nothing but the return of ethical questions and ethical needs. For this reason, sustainable development conceived as institutional reform of the industrial society or a new political trend for rich countries brings no real change in the field, in the life of real people and in the impact of the human species on life and on this planet's resources. This kind of change would remain on the industrial and institutional track. The decision for an ethical turn originates in the feeling that current sustainability policies rely on a limited and finally erroneous understanding of the change *level*. Once we understand that nothing less than the ethical will do, we still have to accept the fact that ethical change is not 'less' but 'more'.

A do-it-yourself ethic for sustainability evades the double-personality syndrome: one ideal self in representation, the discourse's self, and one real and acting self. This is the ethical infra-problem of the present: splitting *representation* (word) and *action* (deed). The infra-ethic of sustainability is self-consistence since the ethical sustainability of the self is the only possible ground for real deeds of the human person. This first tier of ethical awareness leads very naturally to the aggregation of selves for common action, including the management of the commons: local commons, then global commons. It makes a real difference compared to the current sustainable policy, descending on us from the summits all the way down.

> There can be few greater examples of lack of vision in world "leaders" than that, despite their access to the very latest scientific evidence, they have trailed far behind their peoples in recognition of the environmental crisis, which is likely to be the most important political and human issue of the 1990s. [...] Once again it has been ordinary people working through largely voluntary organizations who have acted decisively for human well-being, while the established power structures were either blind to the perils or actively promoting them. (Ekins 1992: 164–5).

Paul Ekins's conclusion is fairly pessimistic, but it can be reinterpreted in the light of Elinor Ostrom's theory of self-governance in 'common pool resource' (CPR) local management. She asserts a fact equivalent to what I call the ethical turn, but in its second phase, the community re-building process:

> What one can observe in the world, however, is that neither the state nor the market is uniformly successful in enabling individuals to sustain long-term, productive use of natural resource systems. Further, communities of individuals have relied on institutions resembling neither the state nor the market to govern some resource systems with reasonable degrees of success over long periods of time. (Ostrom 1990: 1).

Do-it-yourself oriented institutions can do the job for large local commons and in the long run, provided they use the right tools, CPR institutions could also facilitate self-organization, self-governing, monitoring activities and enforcing contracts by oneself. Ostrom has observed these similarities among enduring, self-governing CPR institutions. The most important similarity of all these micro-institutions is their sustainability in itself, meaning here their institutional robustness (Ostrom

1990: 89). Micro-institutional change (founding micro-institutions and managing them), as Ostrom suggests (1990: 139), defines the next step after the ethical turn.

Therefore, for the sake of sustainability research as well as sustainability action, we have an opportunity to construe *ethical change* not as a replacement for political and institutional change, but as the first step toward political and institutional change – provided that we accept the failure of top-down and bureaucratic reform in this domain.

How *deep* is the change we need for a sustainable society to emerge? As deep as the change required for an ethical self to surface. Naess's 'ecosophy' was a search for wisdom and not a science. A bottom-up ecology is 'deep at the bottom', as opposed to 'heavy at the top', which is a common aspect of bureaucratic top-down governance walking on its head. A top-down policy to implement the conclusions of climate science or scientific ecology is an option for the Dark Side, an authoritarian and ideological change.

Sustainability is not directly a challenge to our institutions. It is a challenge to our lifestyle, to our personal lifestyle and to our social lifestyle (institutional, economic and political). They are not 'sustainable' in this evolved sense of the word. They will not last and they do not meet the requirements of human dignity (ethical and ecological dignity). After trying to face the challenge using institutional reform and top-down moral patronizing, we can humbly assume it does not work. The ethical turn offers an alternative.

References

Birnbacher, D. 1994. La Responsabilité envers les générations futures (1988. Verantwortung für zukünftige Generationen. Stuttgart: Reclam). Paris: PUF.

Brown, L.R., Flavin C., Postel S. 1991. Saving the Planet: How to Shape an Environmentally Sustainable Global Economy. New York: Norton.

Diamond, J. Collapse: How Societies Choose to Fail or Succeed. 2005. New York: Viking.

Ekins, P. 1992. A New World Order: Grassroots Movements for Global Change. London and New York: Routledge.

Elgin, D. 1993. Voluntary Simplicit: Toward a Way of Life that is Outwardly Simple, Inwardly Rich. New York: William Morrow.

Gregg, R.B. 1936. The Value of Voluntary Simplicity. Wallingford, Pa.: Pendle Hill. Available at: http://www.soilandhealth.org/03sov/0304spiritpsych/03040 9simplicity/SimplicityFrame.html [accessed: 21 February 2011].

Lovelock, J. 2006. The Revenge of Gaia: Earth's Climate in Crisis and the Fate of Humanity. New York: Basic Books.

Marsh, G.P. 1965. Man and Nature, or, Physical Geography as Modified by Human Action (ed. D. Lowenthal. New York: Scribner, 1864). Harvard University Press.

Naess A. 2005. Selected Works, vol V: Gandhi and Group Conflict: Explorations of Nonviolent Resistance, Satyagraha (1974), revised ed. Dordrecht: Springer.

Nordhaus, T. and Shellenberger, M. 2007. Break Through: From the Death of Environmentalism to the Politics of Possibility (or : Why We Can't Leave Saving the Planet to Environmentalists). New York: Houghton Mifflin.

Ostrom, E. 1990. Governing the Commons: The Evolution of Institutions for Collective Action. Cambridge University Press.

Puech, M. 2010. Développement durable : un avenir à faire soi-même, Paris, Le Pommier, 2010

Richardson, R.D. 1986. Henry Thoreau: A Life of the Mind. California University Press.

Tainter, J.A. 1988. The Collapse of Complex Societies. Cambridge University Press.

Thoreau, H.D. 1854. Walden, or Life in the Woods. Available at: http://www.transcendentalists.com/walden.htm [accessed: 21/02/2011].

Vernadsky, W. 1926 repr. 1997. La Biosphère (translated from the Russian), Paris: Alcan, repr. Diderot Multimédia.

Chapter 3.6

Coping with Climate Change: Social Science and the Case of Multi-site Living

Jørgen Ole Bærenholdt

Introduction

This chapter first discusses an emerging diversity of ways in which social science research in among others human geography and sociology analyse and suggest ways to cope with climate change. Three main ways in which social scientists contribute to coping with climate change can be outlined, illustrated under the headings of Critical geographies, Policy advice and Change in practices. Though simplified in their categorization, the three approaches also imply different figurations of the relation between research and society.

The second part of the chapter exemplifies how coping with climate change has to do with where and how people live. If people increasingly live in multiple places, this raises issues in relation to how people cope with climate change. 'Multi-site living' involves a diversity of possible housing and mobility systems and practices that may either pose a threat or ease efforts to cope with climate change. The case of multi-site living is thus discussed as a complex phenomenon, which could be researched in more debt, especially focusing on Change in practices. The chapter ends with discussing the benefits from researching multi-site living through such an approach, rather than from the first and second approach mentioned in the first part.

The aim of the chapter is to investigate how social science can progress in dealing with climate change in ways that contribute to coping with climate change in practice. While the case of 'multi-site living' is still only a proposal for possible future research, it does offer an example of possible new routes of knowledge making in social sciences.

Critical Geographies

The role of the critical intellectual is the first and in many ways fundamental basis of independent research. For example, it forms the basis of strong traditions of critical geography including radical geography and Nordic *samfundsgeografi* within human geography from the 1970s and onwards, parallel to traditions of critical theory in sociology and around. The emblematic Nestor of critical geography is David Harvey[1] and among his works the central contribution in relation to environmental problems is his *Justice, Nature and the Geography of Difference* (1996). Though this work does not address climate change, it has ways of approaching nature, environmental problems and sustainable development, which implicitly also suggests a position towards climate change, as a process of environmental change, which is always a part of social change, where 'social and political projects are always ecological and environmental projects' (Harvey 2001: 228, mentioning climate change). In his 1996 book, Harvey thus sees 'sustainable development' as a project of ecological modernization. In line with his Marxist tradition, Harvey always sees environmental problems as social problems also involving questions of equity, justice and redistribution. Environmental problems are always problems more for some than for others, and it is therefore also problematic to generalize environmental problems like climate change as the same problem for all (see also Bærenholdt 2007, chapter 9).

There are obvious ways in which the systematic global conflict lines between the rich and the poor world, as they appeared at the little successful COP15 meeting, can be understood on this background. Critical geography also critiques capitalisms interest in making business from climate change. Neil Smith, another prominent critical geographer and student of Harvey, in the after word to his third edition of his famous *Uneven Development* (2008) ironically addresses the ways climate change has become of part of the capitalist imperative for profit and furthermore also 'an excuse for any number of social sins' (2008: 244). Erik Swyngedouw (2010), another student of Harvey, has made a critical inquiry into how climate change contributes to the making of a post-political situation, where social contradictions and antagonisms are hidden under various kinds of populist, post-democratic conditions. Climate change is used to universalize people into 'the people', which is a victim of apocalyptic, uncontrollable forces, beyond struggles over the directions of ecological mere policies or management. Hence, Swyngedouw suggests the necessity against this trend to insist on the possibilities 'for constructing different socio-environmental futures' (Swyngedouw 2010: 228).

Nordic *samfundsgeografi* has much inspiration from critical geography but blended with other inspirations. In the field of nature, biopolitical empowerment and environmental justice, Ari Aukusti Lehtinen of University of Joensuu is one prominent Nordic critical geography thinker, arguing for a more humanist and empowerment oriented position which is more associated with the practices and traditions of people on the margin 'from below', than it is an intellectual critique from 'aside' (Lehtinen 1991, 2006). In his works, Lehtinen is critical to Anglo-American debates in critical geographies that remains within academia: 'How long can we let ourselves be satisfied with the academic progress that is purified to fit

---

1   Honorary doctor (alongside two other key figures of the critical tradition, Oscar Negt and Bob Jessop) of Roskilde University since the university's 25 years anniversary in 1997, and a key-note speaker to geography's 30 year jubilee at Roskilde University in 2004 (see Harvey 2006).

certain discursive dominations, elite lists of key innovators and nicely lengthening records of scientific publications?'(2006: 10). Therefore he suggests working more with a people's geography, taking seriously and intervening in especially the problems of marginalized people. However, when it comes to change in practices, critical approaches, such as those outlined above, may inspire strong critical and political engagements, but their contribution to facilitating new ways of doing things are of limited scope. It lies almost in the nature of the critical position of the intellectual not to intervene in powerful games, such as the politics of climate change, to be discussed in the next section. Elisabeth Shove has commented on the penchant of critical researchers to 'interpret new challenges in terms of existing intellectual positions' (2010: 278), which also implies that the value of critical social theory depends upon not becoming useful for transition processes in relation to climate change. Of course there are examples of both critical and concrete engagement with environmental together with regional development (Hudson 2007) and of concrete action research, for example on flooding, where geographers take an active part in knowledge production and intervention (see discussion in Whatmore 2009), and this shows how critical inspirations may well propel research work engaging in Change in practices (see later section).

Policy Advice

Academics can take another position than the critical intellectual, arguing from 'aside'. They can argue for the need of researcher's practical engagement in giving clever and grounded policy advice. A prominent Nordic example is Norwegian rural sociologist Ottar Brox, who by the way was also a member of parliament in the 1970s. He argues for social science engaging in practical solutions, like marginalisation, poverty and environmental problems, stressing the need to think proactively about possible non-intended consequences (Brox 2007) and has recently published a very practical book on what do to with the climate crisis (Brox 2009). His position as a regional researcher is that there are numerous practical solutions and ways to go, and that these steps do not need to involve a whole revolution of people's way of life. This position offers a somehow rationalistic approach to really deal with the relation between knowledge and action in practice – and it does this more or less 'from above'; from the agenda of the general public and of politicians. It is an enlightenment approach, based on the assumption that publics work (see Førde and Bærenholdt 2007: 297–8).

The iconic British parallel is here world famous sociologist and member of House of Lords, Anthony Giddens, who published his *The Politics of Climate Change* in his own publishing house in 2009, endorsed by Bill Clinton among others. This is an imposing work, collecting information and examples from very many countries (including many Nordic examples). Among many other things, it deals with the geopolitics of petroleum and also suggests a profound 'return to planning'. The whole way of thinking is that of *Realpolitik*, not far from that of Ottar Brox, arguing for ways to do things that will not be so troublesome, as critical intellectuals would argue. In other ways taking the easier ways and short cuts, like using the capitalist imperative to develop new technologies, making the issue get beyond left and right divides, and let political involvement be as much driven by the threat of the peak oil and fossil energy crisis than of the somehow more generalized climate change thing. This is of course in direct opposition to the positions of critical geographers, who would suggest more comprehensive, political and economic, transformation to be necessary.

Such a difference is obvious in geographer Noel Castree's extended review of the book (2010). But this critique has two aspects, which I think should be more clearly separated from each other than apparent in Castree's review: First there is the political critique and irony on the career of Anthony Giddens and his role of supervisor to New Labour in Britain; a critique which also addresses of whole series of other 'political' books by Giddens. This is a political debate, which in fact does not have much to do with research. The other and here more interesting debate, is how books like this one by Giddens (and for example the one by Brox) develop the relation between research and politics. In this debate, Castree has a point to make, if we for a while put aside what was Giddens' intention: Castree sees Giddens' tour de force through enormous amounts of themes making the book superficial, often only summarizing factual knowledge. It is thus a very accessible book, 'but his insights are entirely second-hand and derivative: they do not arise out of scholarly work or deep, sustained thinking' (Castree 2010: 160).

One could of course just accept such policy advice books as what they are: knowledge dissemination from research to a broader audience in order to influence certain political agendas constructively. This is a practice of researchers many international bodies, like UNESCO, often invite researchers to do more of. In that case Castree read the book as if it was something else than it is. However, the other more serious problem, also raised in recent discussions on the Climate Panel's procedures, is what it means to the scientific quality of research, if it was to follow mostly the tracks of policy advice. Worth to mention is here that Brox (2007: 110–11) stresses the crucial role of researcher's autonomy exactly in order to give better policy advice, than can be produced from governmental, or project fund dependent, applied research institutes.

One possible example of how to cope with the dilemmas discussed so far is the work by yet another cross-disciplinary sociologist /social scientist John Urry (2003, 2007, 2008a, 2008b, 2010 and 2009 with Dennis) and associates around Lancaster University and elsewhere[2]. In interesting ways Urry tries to cut across the three simplified approaches, discussed in this chapter. First, his understanding of the dynamics leading to climate change is profoundly based in critical approaches to capitalism, inspired by Karl Marx (Urry 2003, 2007, 2008a, 2010). Second, in

---

2   John Urry is honorary doctor of Roskilde University since 2004, proposed unanimously by three departments of that time, and was a visiting professor 2001–2003 in geography and tourism research.

his work he provocatively suggests a number of future scenarios of how societies could cope with climate change, arguing for the prospects of, the otherwise rather dystopian, Orwellian digital panopticon of control with flows (Urry 2007, 2008b and Dennis and Urry 2009). However this is not concrete policy advice or guidelines in the style of Giddens, but Urry's scenarios in *Mobilities* (2007) is one of the very central references in Giddens (2009: 160). Because, third, the concrete research interest of Urry goes into the systems and practices needed to cope with climate change, where his mobilities engagement has made him focus on the possibilities of a post-car system (Dennis and Urry 2009) not denying and giving up mobility, but finding sustainable ways of mobility. Urry's more distant involvement in politics is embedded in a fundamental assumption about the ways in which the world changes where he seems to believe more in the possibilities offered through various technological, social and cultural paths and trajectories, than what politicians can do. As a researcher he addresses and investigates the complexity and therefore also the contingency of such paths and trajectories, pointing to the fundamental combined unpredictability and irreversibility (Urry 2003, 2004, 2008a) of very many processes.

Change in Practices

The third approach is thus the ways in which researchers engage in more particular engagements in development of new systems and practices, socially and technologically. One way to approach this is via developing *Design Research* developing, reflecting and criss-crossing various ways of combining research with design processes (Simonsen et al. 2010). Here, designing is not only the solution, but also the problem. Many problems of humans of the earth are the result of designs, for example those dependent on and propelling the use of fossil energy. However, this may also point to the need of research interests in exactly design processes, where there is no simple, rationalistic, relation between science and design. In fact the relation between design processes and research is complex, double-bound and diverse. Research-based knowledge do not lead to 'the one and only' right design, and furthermore, much research is in fact design-based, also in the hard sciences, as pointed out in Science and Technology Studies and Actor Network Theory.

It is central to focus on the processes involved in doing design and also to acknowledge that designs emerging from solid analysis of contexts are much more likely to be successful and of lasting value. In fact in our 'mode 2 society' (Novotny et al. 2001), research is increasingly being used, we already live in a society where science is deeply embedded. Depending of the field of particular interest, since researchers always have particular specialities, approaches may differ. In some of the literature the focus is more on transitions, future scenarios, socio-technical systems, path-dependencies, regimes and meta-design (of societies) (Holm et al. 2010, Urry 2003, 2007) and where there is also a particular interest in how certain 'innovations' and 'designs' may disappear in order to deal with climate change (Shove 2010b: 280–82), while other types of research highlight the role of individual practices and dynamic regimes of everyday life involved in climate change (Shove 2010b: 282–84). In the ongoing research work, led by Elisabeth Shove in Lancaster, the focus is on the *'systemic transitions in practice – in patterns of sociability and mobility, and of comfort, cleanliness, food provisioning and leisure'* (Shove 2010a). Combining complex system thinking with everyday practices, new ways of approaching the relation between society and research is sought for (see also Whatmore 2009). The argument is that coping with climate change will involve critical rethinking of our ways of life. It also involves questions of why people do not act and trust in action, which would follow from for example their knowledge on the consequences of daily use of the petroleum driven car (Freudendal-Pedersen 2009, Macnaghten and Urry 1998). Importantly much of such analysis is inspired from social critique in works by for example David Harvey, but they move on to study and interact with people in their everyday life practices. Approaches in various ways manage to study practices and the socio-technical systems, institutions and infrastructure they are embedded in, together. Interests are both in routine and transition, or transformation, and there is an interest to possible very large potentials in even small steps in small worlds. Some approaches are more futuristic in their hopes in possible 'smart' technological solutions (Urry 2008a), while others may give priority to critical questions into why and how, our everyday practices have been ordered and designed (Shove 2010a), opening the agenda for other possible choices.

These concrete approaches shares with the 'policy advice' way of thinking the kind of realism that change processes has to take off and generatively develop out of present systems and practices. They also try to avoid judgement based on generalized distinctions of taste, such as the idea that automobility is bad for everything. Automobility is harmful to climate, environment, safety and heath, but is has also meant various forms of freedom, though some of these are becoming more illusionary with contestation and traffic jam (Freudendal-Pedersen 2009, Urry 2007). Therefore Urry with Dennis (2009) is looking for other, smarter and more sustainable, mobility systems. He sees no realistic possibility nor has he any wishes in suggesting to give up mobility as such. And it is this kind of detailed and specialized interest of particular paths and solutions, which forms the basis for a research-driven interest in ways of coping with climate change, first and foremost in changing systems and practices in order to mitigate climate change. Research, and design, agendas emerging out of such engagements are also those that open for multidisciplinary, interdisciplinary, transdisciplinary or postdisciplinary ways of making researchers from very different specialities work together, through the problems of the particular.

This is the approach now to be exemplified in dealing with the complex phenomenon of 'multi-site living', first introduced in the next section. The example of 'multi-site living' is only meant as a case for how social science can possibly engage with coping with climate change in new concrete ways, taking as a basis the kinds of lay knowledge and practices already available. The case has been selected for a number of reasons: First mobility and climate change is often associated with one another in all too foreseeable ways, repeating to excess more or less populist 'normal' knowledge that human mobility is one of the causes of climate change. To challenge this kind of taken for granted knowledge is therefore one aim. Second human mobility has so far only been studied as a way of coping with climate change as a question of climate refugees, contemporary moving away from for example Pacific islands threatened by rising sea level (see Clarke 2008) or in earlier episodes of human history on earth, where homicides or humans have had to cope with climate change (see Clarke 2010). The other aim is therefore to seek other and 'less vulnerable' examples of how human mobility is or can become a way of coping with climate change.

Multi-site Living and its Implications

It seems that people increasingly inhabit more than one place at the same time; they commute to work on a daily basis, travel for holidays and meet family and friends, or migrate to another place. For periods of their life and, sometimes, within the very same year, many also stay overnight at their workplace for longer periods, in their second home or in the place they migrated from. While such *multi-site living* is not entirely new in many countries, it seems that the *other* place of life has increasingly become within another country – multi-site living has become an international phenomenon. This is in part because of and results in overlaps and convergences between practices of commuting, tourism and migration. Thus, tourists to a place might engage in work there and begin to commute there, residential tourists sometimes become migrants, and migrants may pay return visits to their country of origin and possibly engage in business there. Despite this interconnectedness, social sciences most often either deal with commuting, tourism and migration as separate phenomena or they only discuss these phenomena in their general form without knowing to what extent people practice multi-site living, how they do it and how commuting, tourism and migration intersect.

Multi-site living is paradigmatic in understanding and explaining possible new ways of contemporary social life. While some would argue that migration, tourism and commuting are so divergent forms of practice that a the concept of 'multi-site living' risks to become meaningless, there are two central arguments, of a hypothetical character admittedly, why a common understanding of these forms of mobility connecting to living in more than one place could be useful. The first argument has already been hinted to: There seems to be forms of convergence between migration, tourism and commuting to the extent where even actors themselves cannot make the distinction so that questions about 'political' belonging also emerge, while this suggestion of course needs more research. The second argument has to do with questions of mobility and housing in relation to climate change: In attempting to reduce the use of fossil energy, it matters less how various practices of housing and mobility associated with multi-site living are labelled than more particular questions about how practices could be discussed and changed in order to cope with climate change.

Actual existing practices of multi-site living are thus a condition, a challenge and maybe also a possibility to cope with climate change: Would more sustainable mobility systems, including less frequent trips and sophisticated 'governmobility' (or Urry's digital panopticon), facilitate multi-sited living as a coping strategy; a new kind of system of transhumance allowing people to live in seasonal cycles that are both more healthy to themselves and significantly reduce the need of heating and air condition? Such practices would combine adapting to and mitigating climate change, but they also raise issues such as travelling light, zero-energy pausing/hibernating of dwellings, tempo-spatial organization of work, university terms and other questions of absence-presences. While the prevalence of extensive also leisure based transport (see Holden 2007) and the abundance of people owning or controlling several homes is of course open for a critical approaches to the practices of the elite, there would also be the possibility to think of the migratory bird as a role model! The suggestion is also based on the observation that multi-site living is widely practiced also among non-privileged people, although maybe not to the same extent as elites and with a possibly lower frequency in long trips involving heavy amount of fossil energy. Therefore one could expect that multi-site living among non-affluent people would be especially interesting to study in more detail in terms of everyday practices, politics, types energy use etc.

Research in Multi-site Living

There are some examples of interesting case studies of how people are living on the move and in several places with migration, residential tourism and leisure, work and family practices (Ahmed et al. 2003, Gerrard 2008, Hannerz 1996, O'Reilly 2000, Weaver 2005). And one major contribution comes from Williams and Hall (2000, 2002) offering a number of typologies on relations between tourism and migration.

Research in *commuting* and other similar everyday transport practices are mostly into daily practices, less associated with multi-site living. One example of this is Malene Freudendal-Pedersen's (2009) work on everyday mobility, unfolding the concept of 'structural stories' to understand and explain people's mobility choices. A most dominant structural story is the association of freedom with automobility. Along with other mobility approaches to transport practices in Denmark (Lassen 2009, Thomsen et al. 2005), this work also has a strong emphasis on the social consequences of various forms of mobility. Interestingly, Lassen (2009) is critical to ideas about the happiness among the for ever mobile business travellers; also the life as a privileged 'cosmopolitan' traveller has social costs.

Second, *tourism* has increasingly been studied more concretely as spatial practices with mobilities and places, where tourism also increasingly sweeps out of the realms of the extraordinary and into its various complex connections with the everyday (Bærenholdt et al. 2004, Haldrup and Larsen 2010, Sheller and Urry 2004, Urry 2002). To research in multi-site living, key references come form research into second homes and residential tourism (Duval 2004, Haldrup 2009, Hall and Müller 2004, King et al. 2000, Müller et al. 2004, O'Reilly 2000, 2003, 2009, Willliams et al. 2000, 2004). They study places as performed in practice by tourists. Tourist practice performed here is much about building home and everyday in another place, and with new technologies and flexibility of labour markets, second homes also become places of work (Haldrup 2009: 54, O'Reilly 2009: 135).

Third, *migration* research has led to studies into how labour migration and the practice of multiple 'homes' are embedded in mobile livelihoods, maintaining ties over distances and with the crucial role of remittances (Sørensen and Olwig 2002). In recent years mobility and migration research has increasingly converged, for exampling in research on the role of mobile phones in sustaining network within families, where bread winners are labour migrants (Parrenas 2005, Vertovec 2004a, see Larsen and Urry 2008), research into how refugees establish and maintain even intimate relations over long distances via the Internet (Brekke 2008), or in thinking in new ways about Europe in the light of migration (Fortier 2006). There has been an increasing interest in really understanding the kinds of belonging involved in practicing life in two or more places. Vertovec (2004b) and Skaptadóttir and Wojtynska (2008) have pointed to an emerging 'bifocality', where people take part in belonging, consumption, memories and projects in places far from each other.

Finally, transnational practices raise questions about *citizenship* 'from below' and about how people perform their loyalty and attachment to a *political* unit (Frykman 2008). Mobility research in transport raises questions about sustainable development and policies of regulation (Freudendal-Pedersen 2009, Jensen and Richardson 2007), while tourism research in second homes also address political and planning issues (Hall and Müller 2004). Hall et al. (2004: 22) quotes information that second homes in for example Finland stand for less than 1% of carbon dioxide, phosphorus and nitrogen emissions of the national total, but otherwise the research reviewed above does not deal much with concrete environmental problems and the more exact connections with climate change. While there are reports with both negative and positive considerations on the relation between tourism and climate change, the change in practices approach suggests also investigating the potentials in seeing multi-site living as a way to cope with climate change.

Multi-site Living as Mitigating and Adapting to Climate Change

Admittedly, the idea for this chapter emerged out of double frustration on a cold January day in the unusually cold winter of 2010 in Northern Europe; a winter with snow that began in the second week of the no less frustrating COP15 in Copenhagen, December 2009. Many examples could be approached critically, such as the irony that the smelting of sea ice in the Arctic leads to further exploitation of fossil energy and a whole set of new geopolitical conflicts (as also discussed by

Giddens 2009), while the obvious policy advice would be a moratorium for petrol industries in the Arctic. And many would first approach multi-site living with the same kind of ironic and critical distance; however is this justified? Pastoralist transhumance has been practiced for thousand of years from the Tropics to the Arctic in order to adapt to seasonality of the environment and historically without the use of fossil energy, so are there things to be learned from such traditions (thinking along with ideas from Lehtinen 2005 to Clarke 2010 and Shove 2010b)?

If vacant houses do not need heating or air condition, and if the seasonality of multi-site living is not accompanied with numerous return trips and short visits, the widespread North European retirement migration to the Mediterranean (King et al. 2000, O'Reilly 2000, 2003, 2009, Williams et al. 2000, 2004) could be seen as way of coping with climate change, integrating adaptation and mitigation. Also South European and Asian migrants in Northern Europe take part in such practices. Of course, return trips and short visits to family and friends, which is a crucial part of mobilities' association with social obligations (Larsen and Urry 2008, Urry 2007), is a major challenge. Distant work and education for employed people and school children are major challenges, but with aging populations and also the interest of people of the South to escape probably increasingly hotter summers, increasing migratory practices can be expected whatsoever, simply because most people enjoy the comfort of 'natural' outdoor weather more than heated or air conditioned indoor weather, which has become normal through designs in need of critical questioning.

The migrating crowds of humans in transhumance in for example Europe put a pressure on the amounts of housing space, but if people can stay more outdoor, it also seems that many prefer to distribute their belongings in several places, and therefore (hopefully...) do not need so much housing space in each place. It goes without saying that more sustainable forms of mobility and housing are also in need, but if the idea, which is not abstract but something already widely practiced, would be generalized even further, our normal understandings of societies as made of people living within the territories of nation-states are challenged. This also implies problems for 'normal' understandings of how and to who policy advice could be given.

Conclusion

This chapter has been a thought experiment in two halves. The proposed research interest into what is suggested to be understood as 'multi-site living', is a case of the third 'Change in Practices' approach discussed, highlighting the practices, possible to study, elaborate, experiment and re-design, through a contextual approach to peoples' everyday life. There seems to be little doubt that multi-site living is an irreversible phenomenon, which might have unforeseeable social consequences. Had the chapter only approached this from the Marxist critical geography approach, first mentioned, the phenomenon could easily be investigated along with issues of segregation and gentrification, also in need of being addressed, but few potentials of the trajectory would emerge in such an approach. From the, second, policy advice point of view, multi-site living challenges the most central precondition of policy advice, namely that the policies of nation-states are at the centre of interest (as Giddens 2009 argues for). Admittedly, the European Union is a political reality based on the participation of member-states, and as such it is more than likely that we see the European Union increasingly addressing the good and bad sites of multi-site living in the future. However, more than taking advantages of already emerging practices, such policies will probably start as reactive, 'dealing with the problem' by means of regulation.

Thus turning to the third approach, into change in practice, research into multi-site living is an example of a research interest of a particular kind. The interest is based in contemporary research in mobility and place, why research questions and the specialized qualifications involved starts *neither* from generalized critical stances *nor* from a generalized interest in giving policy advice. Thinking of the potential of multi-site living comes out of research in commuting, tourism and migration, and associated approaches to mobility and place across human geography, sociology and other related traditions. This research builds on the acknowledgement of the practical knowledge accumulated among people on how to design change in practice in time and space. This is a take on production of knowledge in coping with climate change and other environmental problems that integrates the kinds of lay knowledge and everyday experience people already practice (as envisaged by Clarke 2010, Shove 2010 and Whatmore 2009). In this way, the change in practice approach differs from mere policy advice, since it focuses on the emergence of changes in peoples' practices, relevant to understanding process of adaptation to and mitigation of climate change. Change of systems through policies are of course relevant in this context, the point being for research to concentrate on being embedded in understanding and taking part in emerging and potential new practices of change in society. To suggest such a relation between research and society is not to compromise the critical intellectual – nor to compromise the role of politicians – but it seems this is the way to provide better and more useful, contextual and practical knowledge from researchers. It seems that to let research practices built on their knowledge of particular fields, will in effect lead to practically stronger critical and political contributions in designing solutions.

References

Ahmed, S., Castaneda, C., Fortier, A.M. and Sheller, M. (eds) 2003. *Uprooting/regroupings: Questions of Home and Migration*. Oxford: Berg.

Bærenholdt, J.O. 2007. *Coping with Distances*. Oxford: Berghahn.

Bærenholdt, J.O., Haldrup, M., Larsen, J. and Urry, J. 2004. *Performing Tourist Places*. Aldershot: Ashgate.

Brekke, M. 2008. Young refugees in a network society, in *Mobility and Place*, edited by J.O. Bærenholdt and B. Granås. Aldershot: Ashgate, 103–14.

Brox, O. 2007. Anvendelig forskning: Flerfaglig kartlegging av utilsiktede konsekvenser, in *I disciplinenes grenseland: Tverrfaglighet i teori og praksis*, edited by T. Nyseth et al. Bergen: Fagbokforlaget, 101–11.

– 2009. *Klimakrisen: Hva kan vi gjøre?* Oslo: Aschehoug.

Castree, N. 2010. Extended review: The paradoxical professor Giddens: *The Politics of Climate Change. The Sociological Review*, 58(1), 156–62.

Clarke, N. 2008. Climate change: island life in a volatile word', in *Material Geographies: A World in the Making*, edited by N. Clarke et al. London: Sage and The Open University, 7–55.

– 2010. Volatile worlds, vulnerable bodies: Confronting abrupt climate change. *Theory, Culture & Society*, 27 (2–3), 31–53

Dennis, K. and Urry, J. 2009. *After the Car*. Cambridge: Polity.

Duval, D.T. 2004. Mobile migrants: Travel to second homes', in *Tourism, Mobility and Second Homes: Between Elite Landscape and Common Ground*, edited by C.M. Hall and D.K. Müller. Clevedon: Channel View Publications, 87–96.

Førde, A. and Bærenholdt, J.O. 2007. Tverrfaglighet som dialog: Dilemma, potensial og utfordringar i tverrfaglig samfundsforsking, in *I disciplinenes grenseland: Tverrfaglighet i teori og praksis*, edited by T. Nyseth et al. Bergen: Fagbokforlaget, 289–303.

Fortier, A.-M. 2006. The Politics of scaling, timing and embodying: Rethinking the "New Europe". *Mobilities* 1(3), 313–31.

Freudendal-Pedersen, M. 2009. *Mobility in Daily Life: Between Freedom and Unfreedom*. Fairham: Ashgate.

Gerrard, S. 2008. A travelling fishing village: The specific conjunctions of place, in *Mobility and Place*, edited by J.O. Bærenholdt and B. Granås. Aldershot: Ashgate, 75–86.

Giddens, A. 2009. *The Politics of Climate Change*. Cambridge: Polity.

Haldrup, M. 2009. Second homes, in *International Encyclopedia of Human Geography*, edited by R. Kitchen and N. Thrift. Oxford: Elsevier, vol. 10, 50–55.

Haldrup, M. and Larsen, J. 2010. *Tourism, Performance and the Everyday: Consuming the Orient*. London: Routledge.

Hall, C.M. and Müller, D.K. 2004. Introduction: Second home, curse or blessing? Revisited, in *Tourism, Mobility and Second Homes: Between Elite Landscape and Common Ground*, edited by C.M. Hall and D.K. Müller. Clevedon: Channel View Publications, 3–14.

Hannerz, U. 1996. *Transnational Connections: Culture, People, Places*. London: Routledge.

Harvey, D. 1996. *Justice, Nature and the Geography of Difference*. Cambridge, US: Blackwell.

– 2001. *Spaces of Capital: Towards a Critical Geography*. London: Routledge.

– 2006. Neo-liberalism as creative destruction. *Geografiska Annaler, Series B Human Geography*, 88B(2), 145–58.

Holden, E. 2007. *Achieving Sustainable Mobility: Everyday and Leisure-time Travel in the EU*. Aldershot: Ashgate.

Holm, J., Søndergård, B. and Hansen, O.E. 2010. Design and sustainable transition, in *Design Research: Synergies from Interdisciplinary Perspectives*, edited by J. Simonsen et.al. London: Routledge, 123–37.

Hudson, R. 2007. Region and place: Rethinking regional development in the context of global environmental change. *Progress in Human Geography*, 31(6), 827–36.

Jensen, A. and Richardson, T. 2007. New region, new story: Imagining mobile subjects in transnational space. *Space and Polity* 11(2), 137–50.

King, R., Warnes, T. and Williams, A. 2000. *Sunset Lives: British Retirement Migration to the Mediterranean*. Oxford: Berg.

Larsen J. and Urry, J. 2008. Networking in mobile societies, in *Mobility and Place*, edited by J.O. Bærenholdt and B. Granås. Aldershot: Ashgate, 89–101.

Lassen, C. 2009. A life in corridors, in *Aeromobilities*, edited by S. Cwerne et. al. London: Routledge, 177–93.

Lehtinen, A. 1991. *Northern Natures*. Helsinki: Geographical Society of Finland (reprint from *Fennia* 169(1), 57–169).

– 2006. *Postcolonialism, Multitude, and the Politics of Nature*. Lanham: University Press of America.

Macnaghten, P. and Urry, J. 1998. *Contested Natures*. London: Sage.

Müller, D.K., Hall, C.M. and Keen, D. 2004. Second home tourism impact, planning and management, in *Tourism, Mobility and Second Homes: Between Elite Landscape and Common Ground*, edited by C.M. Hall and D.K. Müller. Clevedon: Channel View Publications, 15–34.

Nowotny, H., Scott, P. and Gibbons, M. 2001. *Re-Thinking Science*. Cambridge: Polity.

O'Reilly, K. 2000. *The British on the Costa del Sol: Transnational Identities and Local Communities*. London: Routledge.

– 2003. When is a tourist? *Tourist Studies*, 3(3), 301–17.

– 2009. Hosts and guests, guests and hosts: British residential tourism in Costa del Sol, in *Cultures of Mass Tourism: Doing the Mediterranean in the Age of Banal Mobilities*, edited by P.O. Pons et al. Farnham: Ashgate, 129–42.

Parrenas, R. 2005. Long distance intimacy: Class, gender and intergenerational relations between mothers and children in Filipino transnational families. *Global networks* 5(4), 317–36.

Simonsen, J., Bærenholdt, J.O., Büscher, M. and Scheuer, J.D. (eds) 2010. *Design Research: Synergies from Interdisciplinary Perspectives*. London: Routledge.

Sheller, M. and Urry, J. (eds) 2004. *Tourism Mobilities*. London: Routledge.

Shove, E. 2010a. *Transitions in practice: Climate change and everyday life*. [Online: Lancaster University] Available at http://www.lancs.ac.uk/staff/shove/transitionsinpractice/tip.htm [accessed: 12 March 2010].

– 2010b. Social theory and climate change: Questions often, sometimes and not yet answered. *Theory, Culture & Society*, 27(2–3), 277–88.

Skaptadóttir, U.D. and Wojtynska, A. 2008. Labour migrants negotiating places and engagements, in *Mobility and Place*, edited J.O. Bærenholdt and B. Granås. Aldershot: Ashgate, 115–26.

Smith, N. 2008. *Uneven Development: Nature, Capital and the Production of Space*. 3rd Edition. Athens, Georgia: University of Georgia Press.

Sørensen, N. Nyberg and Olwig, K. Fog (eds) 2002. *Work and Migration: Life and livelihoods in a globalizing world*. London: Routledge.

Swyngedouw, E. 2010. Apocalypse forever? Post-political populism and the spectre of climate change. *Theory, Culture & Society*, 27(2–3), 213–32.

Thomsen, U.T., Nielsen, L. Drewes and Gudmundsson, H. (eds) 2005. *Social Perspectives on Mobility*, Aldershot: Ashgate.

Urry, J. 2002. *The Tourist Gaze*. 2nd edition. London: Sage.

– 2003. *Global Complexity*. Cambridge; Polity.

– 2004. The Complex Spaces of Scandal, in *Space Odysseys*, edited by J.O. Bærenholdt and K. Simonsen. Aldershot: Ashgate, 15–25.

– 2007. *Mobilities*. Cambridge: Polity.

– 2008a. Climate change, travel and complex future. *The British Journal of Sociology*, 59(2), 261–79.

– 2008b. Governance, flows, and the end of the car system. *Global Environmental Change*, 18(3), 343–9.

– 2010. Consuming the Planet to Excess. *Theory, Culture & Society*. 27(2–3), 191–212.

Vertovec, S. 2004a. Cheap calls: The social glue of migrant transnationalism. *Global Networks*, 4(2), 219–24.

– 2004b. Migrant transnationalism and modes of transformation. *International Migration Review*, 38(3), 970–1001.

Weaver, A. 2005. Interactive service work and performative metaphors: The case of the cruise industry, *Tourist Studies,* 5(1), 5–27.

Whatmore, S. 2009. Mapping knowledge controversies: Science, democracy and the redistribution of expertise. *Progress in Human Geography*, 33(5), 587–98.

Willliams, A.M. and Hall, C.M. 2000. Tourism and migration: new relationships between production and consumption. *Tourism Geographies*, 2(1), 5–27.

– (2002) Tourism, migration, circulation and mobility, in *Tourism and Migration: New Relationships between Production and Consumption*, edited by C.M.Hall and A.M. Williams. Dordrect: Kluwer Academic Publishers, 1–52.

Williams, A.M., King, R. and Warnes, T. (A). 2004. British second homes in Southern Europe: Shifting nodes in the scapes and flows of migration and tourism, in *Tourism, Mobility and Second Homes: Between Elite Landscape and Common Ground*, edited by C.M. Hall and D.K. Müller. Clevedon: Channel View Publications, 97–112.

Williams, A.M., King, R., Warnes, A. and Patterson, G. 2000. Tourism and international retirement migration: new forms of an old relationship in southern Europe. *Tourism Geographies*, 2(1), 28–49.

Section 4
Communication

Introduction to Part 4:
Climate Change and Communication

Pernille Almlund

In this communication section, researchers within the field of communication focus on the importance of both the communication of climate and climate change, as well as the importance of analysing climate communication. The next five chapters present various case studies, which underline how analysis is just as important as action. They demonstrate how the analysis of climate communication can improve action and provide us with insight into understanding climate change within different systems, locations and uses.

The following communication section focuses on the importance of communication in regards to maintaining knowledge about the challenges posed by climate change and how to communicate and perform in a better way. The section begins with a chapter that examines how climate communication can be altered and improved in Bangladesh based upon an analysis of climate communication in the country. The next tree chapters then go on to analyse specific climate events, namely: documentary films about climate change, the legal event *Earth Hour* and the illegal direct action called *Shut it down*, which attempted to close down Amager Power Station, a Danish power plant. Finally, the last chapter analyses climate communication amongst climate spokespeople from eight different political parties in the Danish Parliament in order to understand how climate communication actually takes form.

In response to the research gap between investigations concentrating on developed countries vulnerability to climate changes in comparison to developing countries, Irene Neverla, Corinna Lüthje and Shameem Mahmud provide an examination of challenges to mass media climate change communication in Bangladesh. By examining the mediatisation of climate change in Bangladesh in their chapter entitled, "Challenges to Climate Change Communication through Mass Media in Bangladesh: A Developing Country Perspective", the authors identify two sets of challenges; first, *issue constraints* and second, *journalistic constraints*. They underline how socio-economic and geophysical conditions, paired with the media system and journalistic culture, make climate change communication in the media more difficult in developing countries like Bangladesh, than in more developed societies. The examination is framed by a theoretical analysis of communicating climate change that looks at the dual issues of 'risk' and 'science'.

Embracing a discursive analytical approach, Gabriela Ramirez Galindo illuminates us how different types of climate change documentaries are portrayed in the UK media, in her chapter "Could Films Help to Sae the World from Climate Change? A Discourse Exploration of two climate change documentary films and an analysis of their impact on the UK printed media". The chapter analyzes the top-down documentary, *An Inconvenient Truth* (2006) by Al Gore and the bottom-up film, *The Age of Stupid* (2009) which comes from a diversity of first-hand experiences. Gabriela Ramirez Galindo shows how the presence of Al Gore in *An Inconvenient Truth* elevates the film to a controversial climate change and political reference in the UK media. She also demonstrates how *The Age of Stupid* achieves a positive consensual coverage, but with a shorter and more entertainment-centred focus. Both films ensured climate change presence in the media, but not always with the same echo for calls of urgent action.

In his chapter, "Communicating the Political Act of Switching off the Light: Mediating Citizen Action through 'boundary acts' in the Earth Hour and Vote Earth global media events", Paul McIlvenny examines the discourses and mediated actions through which citizens can take part in a global climate media event called *Earth Hour* (2009). During *Earth Hour*, ordinary people were called upon to switch off their lights as a collective symbolic material act in recognition of climate change. In the sister campaign, called *Vote Earth*, the discourse of representative democracy was deployed to constitute an imaginary global electorate, where people could vote for the Earth by switching off their lights. Paul McIlvenny focuses on the intense drive to visualise and politicise the simple act of switching off the lights and the consequences of such a 'boundary act. He then moves on to examine the discursive 'memory work' employed to archive and memorialise the hour on outlets such as YouTube. Lastly, he focuses on resistance to the circulation of this 'new' mediated discourse in video blogs.

In the chapter, "Climate Activism and the Mass Media: Potentially a politically challenging debate," Anders Danielsen examines how activist's critical messages of climate change during the illegal protest *Shut it down,* gained a voice in the Danish mass media debate. Anders Danielsen argues that although the debate could have been a political challenge, it was far from one for three reasons. First, news coverage was chiefly focused on the direct action, second, the recipient of the critique, the state-owned Swedish power company, Vattenfall, did not perceive a conflict of interest between activists and the company and third, the activists were generally unable to articulate critique with a broader structural perspective in the mass media.

With the ambition to identify how politicians in the Danish Parliament understand and communicate climate, Pernille Almlund examines politicians communication about climate through political negotiations in the parliament, as well as through face-to-face interviews with spokespeople from the different parties represented in the Danish Parliament. She demonstrates how the discussions are dominated by the issues of *cost for society* and the differentiation between mitigation and adaptation. Within this frame, arguments are led by political opinions and despite an overall agreement about the importance of taking action on climate change, for example at the COP15, the communication of climate still remains a politically charged discussion. These results show how the communication of climate is more an internal political issue located within the political system than of the handling of climate changes. The investigation is carried out within a theoretical framework based upon the System theory of Niklas Luhmann and the Praxis theory of Pierre Bourdieu.

All five chapters within the following section are based on very different communicative actions; this variety underlines the importance of communicative investigations. The chapters shed light upon the role of: the media, legal and illegal climate events, climate films, and political climate communication. In this way, all five chapters contribute to an increase in the understanding of climate change, as well as to the importance of studying climate change through a communicative perspective.

Enjoy your reading.

Chapter 4.1

Challenges to Climate Change Communication Through Mass Media in Bangladesh: a Developing Country Perspective

Irene Neverla, Corinna Lüthje and Shameem Mahmud

Introduction

Possible anthropogenic climate change is widely discussed and investigated as one of the most concerning global risks of our time – perhaps of all time. It first appeared in the works of climatologists and other natural scientists who predicted likelihood of climate change due to human-induced increases in the atmospheric concentration of green house gases (Houghton 2004, Ward 2003, Bernstein et al. 2007). In the beginning, the debates on 'green house gases', 'global warming', and 'climate change' were limited within the inner circles of science and scientific literature. Since the late 1980s, a number of extreme weather events across the world coupled with increased scientific research on global warming and its gradual politicisation have set climate change as an important agenda for news media and politics. By the end of the 20th century, global warming and climate change have become "grand narratives" (Neverla 2008) for sciences, media, politics, business, and to some extent for public in some of the advanced societies. Climate change issues continued to receive increased importance in media and politics in the first decade of the 21st century. Again, this decade was marked by a number of extreme weather events (e.g., European heat wave in 2003, Indian Ocean earthquake and tsunami in 2004, and Hurricane Katrina in 2005); high profile assessments on future climate by the Intergovernmental Panel on Climate Change (IPCC); the annual Conference of the Parties (COP); the 2007 Nobel Peace Prize jointly awarded to the IPCC and former US vice president Al Gore; and most importantly continual politicisation of climate science. No doubt climate change will continue to receive attention and interest in the decades coming. With these changes in mind, it suffices to say that climate change is no longer a mere environmental and scientific concern, rather a political and economic issue at national and global levels. Furthermore, climate change is not only an issue of 'politicisation' and 'mediatisation' in developed countries, but also for countries in the global South (Schäfer et al. 2011, Boykoff & Mansfield 2010, Billet 2010). Politicians and environmental activists in the developing countries are voicing concerns about anthropogenic climate change in national and global platforms more frequently than ever before. They are also taking initiatives for mainstreaming climate change in national development policies, and to improve public understanding about risks of climate change (e.g., BCCSAP 2009, NAPCC 2009).

Public understanding is an integral part of risk management process as Leiserowitz (2006: 45) puts it: "Public risk perceptions can fundamentally compel or constrain political, economic and social action to address particular risks". Communication (both mass and interpersonal), on the other hand, holds important roles in the process of constructing public understanding of risk (Neverla & Schäfer 2012, Smith 2005, Fischhoff 1995). Accordingly, media coverage of risk has become an important subject of risk communication studies which has been developed in the last three decades or so. However, compared to many other socio-economic, environmental and scientific risks, climate change is difficult for media to communicate precisely and effectively. The volume of research on climate change media coverage and public perceptions revealed that understanding of the causes and consequences of global warming by the lay public is quite different from that of scientists and policy experts (Lorenzoni & Pidgeon 2006, Stamm et al. 2000, Kempton 1991). Storch (2009) further pointed out two competing types of knowledge – *the scientific construct* and *the cultural construct* of anthropogenic climate change. In the later case, media play the crucial role by building a bridge between scientists and public. Stamm and colleagues (Stamm et al. 2000), in fact, identified limited understanding of public about global climate change as a communication problem.

Social scientists quickly responded to this problem and developed a considerably large volume of research in the areas of media portrayal of climate change (e.g., Trumbo 1996, Carvalho & Burgess 2005, Olausson 2009, Dirikx & Gelders 2010); relations between public understanding and media representations (e.g., Bord et al. 1998, Stamm et al. 2000, Bell 1994); and challenges to climate change communication through mass media (e.g., Wilson 2000, Boykoff & Boykoff 2007, Filho 2008, Carvalho 2008). One salient feature of this wide variety of studies is that almost all of them took media samples and people in developed countries as the cases. On contrary, little is known about the situation in developing countries which are more vulnerable to climate change mainly due to socio-economic conditions (Nagel et al. 2008) and limited hazard management capacities (Mirza 2003). In response to such a noticeable research gap, this chapter sets out – first, to provide a theoretical analysis of communicating climate change as a dual issue of 'risk' and 'science'. Second, we examine challenges to climate change communication through mass media in a developing country. We have taken Bangladesh as the case for specific reasons. First and foremost, the country is often referred in scientific literature and political discussions as one of the worst victims of man-made climate change. Second, in recent years there has been an increasing importance of climate change in both policy and media discourses of Bangladesh taking it as a threat to its development. This chapter begins with a brief outline of geophysical and socio-economic conditions of Bangladesh which will contextualise the subsequent discussions on theoretical concepts of risk and science communication as well as climate change communication challenges for Bangladeshi mass media.

Bangladesh and Climate Change

Bangladesh is located in between two different geophysical conditions with Himalayas to the North and Bay of Bengal to the South. This particular location situates Bangladesh in proximity to frequent natural hazards that together with socio-economic and demographic conditions often turn into disasters. Tibetan glaciers are sources of several largest rivers in the world including the GBM (Ganges-Brahmaputra-Meghna) basin which has created Bangladesh as the world's largest delta after years of siltation. Several hundred rivers and their tributaries have crisscrossed the country before falling into the Bay of Bengal. Water flows from upstream Himalayan glaciers through these rivers and heavy rains during the monsoon season often cause severe floods in Bangladesh affecting millions of lives. On the other hand, waters of Bay of Bengal and the adjoining North Indian Ocean are sources of tropical cyclones, storm surges and coastal erosion (Ali 1999). The 2004 Disaster Risk Index (DRI) developed by the UNDP listed Bangladesh as most vulnerable to tropical cyclones and floods (UNDP 2004). According to the DRI 2004, Bangladesh accounts for more than 60 per cent of all deaths associated with tropical cyclones in between 1980 and 2000 (UNDP 2004: 37). In Bangladesh, climate change vulnerabilities are results of high physical exposure to geo-hazards of a large population, particularly 35 million people who lives in the coastal zones (UNDP 2004: 37). Bangladesh, with an estimated population of 160 million living in an area of 147,570 sq. km., is world's one of the most densely populated country. Its per capita income is US$ 580 and 40 per cent of the population lives below the national poverty line (World Bank 2009). It has an agrarian economy which accommodates the major labour force. The adult literacy rate is 55 per cent. All these factors together – geophysical conditions, population size, high level of poverty, and dependency on agriculture – make Bangladesh more vulnerable to geo-hazards. By geo-hazards, we mean particular forms of human-nature interaction and one precondition is mankind's settlement in an endangered region despite potential threat. Any changes in the global climate will only increase those threats.

Communicating Climate Change: Theoretical Concepts

Communication of climate change through mass media involves at least two salient features. First, it is about communication of risk, and second communication of science. In general, scientific information is relatively unattractive for media coverage compared to news about politics, crime, sports, entertainment and so on. However, newsworthiness of science increases if it involves threat to human life (Weingart et al. 2000). We can argue that both media and their publics may have less interest about the issues and events which deal with the complexity of climate science but the same issues can attain significant importance once they convey personal or societal threats. Accordingly, theoretical concepts of climate change communication intersect the concepts of 'risk communication' and 'science communication' – two important streams of communication studies. Climate change, in fact, brings these two streams together.

Climate Change as Risk Communication

The term 'risk communication' is more a recent construct which first appeared in the literature in 1984 as a follow-up of risk perception research (Leiss 1996). According to Gurabardhi and colleagues (2004), the basic idea of risk research in the 1980s and 1990s was that understanding public perception of a risk event or issue would enable researchers to develop more effective communication models and strategies, and effective use of those strategies will reduce gap between expert and public understandings of risk. Scholars also studied media portrayal of risk issues and events to understand trends and nature of coverage that may influence public perceptions. As this study is mainly concerned with communication of climate change through mass media, we want to shed some light on the studies of media coverage of risk. For this, we rely on Kitzinger (1999) who provided a brief, but brilliant meta-analysis of scientific literature dealt with risk coverage in media. Kitzinger (1999) argued that this is unlikely to generate any universal theory which would describe 'how and what' determine media coverage of risk. Rather, transmission of risk through news media is a complex process which might be influenced by many factors. Renn (1991: 307) apparently identified three important factors: first, the issue itself; second, the institutional context; and third, the political salience of the issue. Taking on board the studies on media coverage on risk, we can argue that climate change risk coverage in the media depends on the nature and extent of risk, together with its politicisation, media systems and journalistic cultures; socio-economic and geographic conditions; and most importantly relevance of the issue to the people. That is, possible impacts of climate change in the locality of the media.

For the purposes of ensuing discussion on climate change media communication as a risk issue, we present here some common features of risk coverage in media. First, coverage of risk tends to be event oriented rather than issue oriented (Kristiansen 1988, cited in Kitzinger 1999: 62). Media coverage of an issue usually lasts for a longer period than coverage of an event which comes and fades away quickly once new event appears. This proposition supports trends of ups and downs of climate change media coverage instead of a consistent visibility that was documented in the longitudinal media content analysis of Ungar (1992), Mazur and Lee (1993), Trumbo (1996), McComas and Shanahan (1999), and McManus (2000), among others. One of the main arguments of this genre of research was based on Downs's (1972) issue attention cycle which argued that media attention towards a particular issue move through a five-step process in which interest moves to the pick and then gradually declines. Media attention to global warming and climate change issues also depends on some critical moments such as occurrences of extreme weather events, important global negotiations of climate change or publication of new scientific report (Carvalho 2005). We can argued that climate change is not an issue of daily coverage. Increasing competition of other new issues makes it difficult for the news media to provide regular coverage for something which is largely invisible and abstract to people. Global climate change, to some extent, is such an issue.

Secondly, research on media reporting of science suggests that journalists look for science stories 'with a human angle' (Hansen 1994: 114) that is to suggest that media would frame climate change science and associated risks from a human angle and would use verbal, graphical and visual metaphors to stimulate emotional feelings of audiences by relating it to their daily life. In Germany, the picture of the half-submerged Cologne cathedral has become the icon of the threat of global climate change for newspapers to illustrate the issue as a catastrophe (Weingart et al. 2000). Similar visual metaphors (e.g., polar bears on melting ice) of climate change are very common in media reporting of global warming across the world.

Thirdly, conflict and blame are key criteria in media coverage of risk (Kitzinger 1999). The so-called 'news values' set media interest about 'conflict' and 'controversy' more than 'consensus' about an issue. Media policymakers contend that people want to know more about 'who are disagreed, on what, and why' than news about 'who are agreed on what'. Journalists want to identify conflicting parties on the causes and consequences of risk, and also 'something' or 'somebody' to blame (Sandman et al. 1987, cited in Kitzinger 1999). Climate change media coverage research supports this trend when both 'conflict' and 'responsibility' appeared as major frames of media coverage (Boykoff & Boykoff 2004, 2007, Olausson 2009, Billet 2010).

Fourth, despite ubiquitous discussions about globalisation of news, so far journalism is best practiced at local level although there are trends of internationalisation of news organisations. Media outlets apply different tools for domestication of global risks. Neverla (2007) argued that coverage of environmental risk issues in national media are not synchronous and consensual, rather asynchronous and controversial which stress the own national point of view.

Finally, journalists' 'mental maps' are important determinant of the way they frame news story (Dunwoody & Griffin 1994, cited in Kitzinger 1999). Journalists' background knowledge and attitude towards the risk issue largely influence the extent and quality of coverage. This trend is particularly true for coverage of an issue which involves complex science and scientific methods.

Climate Change as Science Communication

In general, two models are discussed concerning communication of science to the public – the public understanding of science (PUS) and the public engagement with science (PES). Bauer's (2009) study on evolution of the public discourse on science, however, finds three paradigms that developed since 1960s – science literacy paradigm (1960s–1980s), the PUS paradigm (1985–1990s); and the science-in-society paradigm (1990s to present). According to Bauer (2009), both PUS and science-in-society paradigms are rooted in the scientific literacy perspective that developed in many advanced and developing countries during the 1960s and 1970s with an objective to increase public knowledge of science. In principle, the idea of scientific literacy attributes a knowledge deficit to the public and call for knowledge transfer from the top (scientists and policymakers) to the down (lay people) to fill the gap. Numerous attempts were taken to increase public knowledge of science in the 1970s through formal and informal science education programmes involving the media (Bauer 2009). The model did not work in most cases. Instead, it has been found that public distrust and scepticism about science were increasing (Sturgis & Allum 2004). The 1985 report of the Royal Society of London influenced largely a transition from literacy paradigm towards the PUS model. According to Bauer (2009), the PUS model was designed not only to increase public knowledge, but also to change public attitude to science. But, this model also received criticism on the grounds that public deficit of knowledge was considered as main problem and it had been prescribed to fill the deficit pouring information from scientists. As documented by Bauer (2009: 231), the operational axiom of 'the more you know, the more you love it' did not work in all cases. He argued the PUS model might be correct in the context of a developing and industrial society while in the knowledge intensive post-industrial context, this is no longer the case. We, however, disagree with Bauer's view on developing countries as the PUS has been prescribed for developing countries in the same linear and top-down fashion as development communication was advocated in the 1970s. In the PUS model, lay people were seen as obedient receivers and respondents of information getting from policymakers, mainly through mass media. From a communication perspective, the PUS considers the knowledge flow in a linear model – from those who know more to those who are perceived to have a deficit of knowledge. But, communication *per se* is grounded in dialogical model, albeit flow of information from different participants in the network of dialogue is not equal. In the case of climate change, lay people may have limited understanding of scientific complexities, but they possess relevant and important knowledge of their own socio-geographic conditions as well as century old coping strategies with nature and natural hazards. It has been also found that the audiences often act as critical consumers and even producers of science information for their own purposes rather than the previous idea of passive receivers (Irwin & Wynne 1996). Accordingly, the PES model is suggested acknowledging lay people's stake in the process. The main assumption of the PES model is that both publics and scientists have expertise, valuable perspectives, and knowledge to the development of science and its application in society (Leshner 2003). In modern societies, this is impractical to imagine direct communication between scientists and publics at large extent. Here lies the importance of news media, which can create the space where both scientists and publics can come together to share their respective perspectives.

Challenges to Climate Change Communication: Issue and Journalistic Constraints

The second part of this chapter discusses some of the common challenges to communicate climate change through news media in Bangladesh. We argue for two sets of challenges – First, the *issue constraints* – these are the challenges mainly originated from the climate change issue itself. Examples include the politicisation of climate change issues, and the uncertainty of climate science. The second genre of challenges are more linked to news-making processes and can be labelled as the *journalistic constraints*, which include dependency on foreign media; problems in sources; and journalists' knowledge. One may argue that such constraints to media communication of climate change are universal, particularly the issue constraints which in general shape the trend and nature of media coverage across the world. However, we argue that the implications are quite dissimilar. Specially, the second set of constraints multiplies the challenges of climate change communication in developing countries because of their socio-economic conditions, media systems and journalistic culture. We will present the Bangladeshi case here.

Issue Constraints

Politicisation of climate change

Climate change is now a mainstream political issue of national and global politics (Giddens 2008, 2009, Haldén 2007). The politicisation process, however, began since emergence of climate change as an important policy agenda in the late 1980s, particularly in the US, where conservative think tank questioned first about 'validity of global warming science' and then evidence of 'climate change' predictions. The global warming deniers successfully influenced climate change policy making in the US which included its rejection to support the global warming treaty of 1997 known as the Kyoto protocol (McCright & Dunlap 2003, 2010). On the other hand, politicians by and large in Europe and Least Developed Countries (LDCs) see climate change as a real threat and emphasise on initiatives to mitigate its adverse impacts. The third group – countries known as emerging economies that include, for example, India and China - agree on the threat of climate change, but engage in bitter conflict with industrialised nations on the question of reducing their own carbon emission as mitigation actions. All these three groups – industrialised or developed countries; emerging economies; and the LDCs - are in fact engaged in political conflicts that have been exposed noticeably in different international policy summits. Other categorisation of climate change politicisation is more generic and two groups are mostly visible in the scene – alarmists and sceptics. According to Giddens (2008), alarmists are viewed as mainstream of climate science knowledge and mainly represented by the IPCC. Alarmists provide a consensus about anthropogenic causes of the extent and dangers of global warming in the form of climate change. On contrary, climate change sceptics, although small in number, put forward the argument that science of manmade global warming is not settled. Some of these scientists as well as politicians even argue that climate change is a natural phenomenon and green house gases do not have any relation to it (Giddens 2008). Two more groups can be identified, which actually originated from alarmists and sceptics blocs. Some of the alarmists, whom Giddens (2008) calls radicals, think climate change poses even greater, and more urgent threats than is ordinarily acknowledged. Supporters of this group see climate change not as a future possibility, but a present reality since they argue that many parts of the world have already been affected by changing climate. We identify another group who may be called 'pragmatic'. The pragmatics believe that global warming is true and increased emissions of green house gases are creating the problem. But, they are unconvinced of alarmists' scientific predictions about future climatic changes which they believe is often exaggerated in media.

These diversified positions of politicians, climate scientists and other stakeholders have important implications for communication of climate change to the public. First, various stakeholders (e.g., scientists, politicians, corporate business members, NGOs, environmental activists, civic groups, celebrities, etc.) have emerged in the political scene of climate change debate. Most of them are alarmists and some are sceptics. These stakeholders, with their respective positions approach the media to manipulate media power for public attention. As a result, the political polarisation is often extended to the public. McCright and Dunlap (2011), for example, found that people in the US are divided on their beliefs with scientific consensus about global warming according to party policies of liberals (Democrats) and conservatives (Republicans). Second implication is more general as through the politicisation process scientists actually lost their definitional control of the debate in media, which is now shaped and controlled by politicians (Trumbo 1996) and environmental activists of NGOs (Mahmud 2010).

Content analysis of Bangladeshi newspaper coverage of climate change revealed that radical government policymakers and NGOs are main sources of climate change issues (Mahmud 2010). These people appeared as credible, authoritative and legitimate definers of reality of climate change science. The role of NGOs in the mediatisation of climate change demands special attention in Bangladeshi context. NGOs are important actors of socio-economic development in Bangladesh which pioneered micro credit as poverty reduction tool and produced couple of world's largest NGOs (e.g., BRAC and Grameen Bank). They work on issues from poverty reduction to women empowerment, education, human rights, natural hazard management, and of course climate change is a recent priority. NGOs in developing countries often work as issue entrepreneurs and this was particularly evidenced in Bangladesh during the COP15 summit in December 2009 when a large number of NGOs worked to generate media events both in the country and at the conference venue in Copenhagen about adverse impacts of climate change in Bangladesh. Their communication campaigns, financed by the government and donor agencies, portrayed climate change in Bangladesh as 'already visible' and claimed that country's recent extreme weather incidents were evidences of changing climate. This is hardly any climatologist who explains complex climate science in Bangladeshi media, but the NGO activists who are of course biased to their narrow political agenda (Mahmud 2010). We find similar trends in Smith's (2005: 1473) work who argues that NGOs claims are not fixed, but often 'opportunistic and innovative in ways that satisfy news needs and practices'.

'Uncertainty' vs. 'absolute certainty'

If uncertainty of climate science and politics of uncertainty were major challenges for media in developed countries to deal with the issue, media in developing countries such as Bangladesh are completely on the reverse side. News media in Bangladesh and India hardly raise any question about uncertainty of climate change science where 'absolute certainty' is the dominant frame (Billet 2010, Mahmud 2010). 'Absolute certainty' demands an explanation of 'uncertainty' of climate science debate at least for the sake of a comparative scrutiny of the perspectives of climate science coverage between developed and developing countries.

The term 'uncertainty' has different meanings for science, policymakers, and public (Briscoe 2004) who together form a 'triple interface' (Boykoff 2010: 399) mainly connected through mass media. For science, uncertainty doesn't have a negative connotation, but an intrinsic element of any scientific inquiry (Briscoe 2004). In principle, science is a process to reach the point of certainty, but the paradox is that every scientific discovery towards that goal of certainty creates new uncertainty – new questions to answer. Uncertainty often breeds indecisions for policymakers in taking actions, which should depend on concrete and consensual scientific discoveries. For this uncertainty debate, policy discussions of climate change are always centred on two options – one calls for immediate action (such as cutting green house gases emissions); and the other calls for 'go slow' and 'wait-to-learn' approach (Webster 2002).

What role media play in this political scene of uncertainty? As we have already seen media normally don't like uncertainty as newsworthy, but prefer controversy and conflict surrounding the uncertainty. In the context of an issue with any degree of controversy and uncertainty, Smith (2005) argues journalists take a position in between the lines and balance the coverage. Accordingly, arguments of mainstream climate scientists are balanced with a small number of sceptics. The ultimate result is a biased coverage (Boykoff & Boykoff 2004). Thus, we can draw a linear equation of relations between 'uncertainty in science' and 'journalistic norms', that is, news media first attempt to balance the controversy by bringing both sides of the coin, which eventually results in bias coverage as media balance majority with minority. This is the case what happened in the climate change news stories since the beginning particularly in the US media and British tabloids (Boykoff & Boykoff 2004, Antilla 2005, Dispensa & Brulle 2003). Recent coverage of the US media (between 2003 and 2006), however, showed a paradigm shift of the quality of coverage as the media are increasingly emphasizing on the scientific consensus of the anthropogenic causes of climate change (Boykoff 2007, Kuha 2009). However, trends of 'uncertain climate change' coverage are still prevalent in the British tabloids (Boykoff & Mansfield 2008). Furthermore, systematic media campaigns by global warming sceptics from business and politics contributed to construct an uncertain reality of climate science in media discourses (Antilla 2005, McCright & Dunlap 2011). The sceptics used public relations and lobbying techniques to propagate the debate through media that scientific claims of anthropogenic climate change are nothing but 'a hoax' (Antilla 2005).

In a clear contrast to western media coverage of climate change science, Mahmud (2010) finds that almost all articles in Bangladeshi newspapers frame climate change as anthropogenic and put the blame on both developed and emerging economies for creating the problems of climate change. Bangladeshi newspapers, taking a radical position, framed climate change science not only as certain, but as 'it is happening now'. For example, the leading daily newspaper of the country – *Daily Prothom Alo* ran a regular slogan during the COP15 that read, "If Bangladesh survives, the world will survive" (translated from Bengali). This is also interesting to note that none of the two major newspapers that Mahmud (2010) examined gave any coverage to the so-called 'climate gate' issue that was about leaked e-mails of scientists from the University of East Anglia's Climate Research Unit. The climate gate issue was widely covered and discussed in western media on the eve of the COP15 initiating new controversy about validity of global warming science. We argue that media coverage of climate change in Bangladesh is based around the frames of 'risk' and 'responsibility' and the narrative is purely nationalistic. Similar results were found in Indian newspaper coverage (Billet 2010). Importantly, Bangladeshi newspapers overwhelmingly linked natural hazards (e.g., cyclones, floods and droughts) with climate change and frequently used the metaphor that much of Bangladesh would inundate as a result of sea level rise (Mahmud 2010). We also argue that such 'absolutist view' is another problem of communication of the issue to the public. Overwhelming certainty arguments of climate change and the claimed link to natural hazards may give a false impression of the situation to the public who may want to find evidence in their surroundings about the claims made in media. More importantly, such contention about climate science mainly came from radical politicians and environmental activists, not from the scientists.

Journalistic Constrains

Dependency on foreign news media

Non-western media often reproduce a considerable amount of climate change stories from European and North American news media (Boykoff & Roberts 2007) and international wire services (Antilla 2005). This feature of climate change coverage is particularly true for the developing countries (Harbinson et al. 2006) and Bangladesh is not an exception. A study on origin of climate change news stories in Bangladeshi newspapers during the COP15 showed that international news agencies and foreign media (mainly Western) were sources of 42 per cent of all articles (Mahmud 2010). This clearly suggests a dependency on foreign media for climate change news stories that is mainly because of limited institutional capacity of Bangladeshi media to cover events at global stage. In this context, foreign media ultimately play a key role in shaping the ways in which climate change stories are framed and discussed. The ultimate result of this dependency is ignorance of local perspective of the problem. Shanahan (2009) depicts a vivid picture of capacity of developing countries' news media to cover global climate change. He writes:

> "Yet of 1,500 journalists who applied to attend the UNFCCC summit in Bali in December 2007, just 9 percent were from non-industrialised countries other than the host Indonesia (Fahn, 2008a). A much smaller percentage actually travelled to Bali and for nearly the entire UN list of 50 Least Developed Countries, there was zero media representation." (Shanahan 2009: 146).

In such a situation there is limited alternative for news media in developing countries other than depending on wire services and foreign media, thanks to the Internet that made it easer nowadays to get those stories quickly and free of cost. The central premise of our argument is that while coverage of climate change in non-industrialised countries like in Bangladesh has increased in recent years, the quantity of reports does not match with quality of coverage. Furthermore, quantity of coverage only increases when there is a high-profile global event like the COP or if there is any important scientific publication, like the IPCC reports.

Now the relevant question is what are the implications of such foreign dependency of climate change coverage? Lets have a look on one important aspect of climate change and global warming reporting, that is, finding solutions to global warming problem. In general, two types of solutions are discussed in scientific and policy discussions– mitigation and adaptation. Mitigation strategy has been defined as interventions for reducing the emissions of green house gases (Koivurova et al. 2009). On the other hand, the IPCC defines adaptation as "adjustment in natural or human systems in response to actual or expected climatic stimuli or their effect, which moderates harm or exploits beneficial opportunities" (Bernstein et al. 2007). Mitigation solutions are more relevant for the industrialised countries which are mainly responsible for emissions of green house gases and accordingly mitigation is widely discussed in policy and public agendas there. On contrary, as a potential victim of global warming, adaptation strategies should be communicated to the public in Bangladesh more adequately than mitigation options. However, we find Bangladeshi newspapers provided more coverage on mitigation than adaptation as solution to the problem. Mahmud (2010) found that mitigation strategies were referred in 54 per cent of the news articles that discussed about solutions to climate change while adaptation options were discussed in only 25 per cent articles, and rest 21 per cent articles provided coverage on both mitigation and adaptation. This is not because Bangladeshi newspapers preferred mitigation as more important for their readers than adaptation. Rather, the reason is overwhelming dependency on western news sources. As Mahmud (2010) further revealed that locally generated news articles comparatively emphasised more on adaptation options as compared to articles from foreign sources.

Problems in sources and news actors

Actors in the news articles are important definers of issue agenda. By actors, we mean the people who are either quoted in the articles or from whom journalists derive information or sources of news. In Bangladeshi newspapers, we find an overwhelming presence of government policymakers, NGO activists and UN officials as dominant sources of climate change news stories who all together comprise around 70 per cent of all sources cited in the news stories (Mahmud 2010). Scientists were poorly represented and lay people were almost ignored in Bangladeshi newspaper coverage of climate change. Only 10 per cent of the actors cited in the news articles were scientists (almost all of them were foreign scientists) while less than 6 per cent were lay people (Mahmud 2010). This is clear that Bangladeshi news media have failed to bring different perspectives of climate change issue by giving main emphasis on policymakers when two other important pillars (science and public) of the triple interface of 'science-policy-people' were underrepresented in the coverage. Trumbo's (1996) longitudinal study on the US media coverage finds the similar trend when emphasis of the coverage shifts away from a presentation of the issue in terms of its causes and problematic nature and toward a presentation more grounded in political debate. One implication of this trend of coverage is failure of any attempt of communicating science to public, no matter whether it is designed as 'public understanding of science' or 'public engagement with science' model.

Another salient feature of Bangladeshi media coverage of climate change is overt negligence of rural areas, more precisely the areas which are more vulnerable to climate change risks. Mahmud (2010) finds only 3 per cent news articles appeared in the newspapers which were written after visiting vulnerable areas in rural Bangladesh. That is to say that urban centric journalism which produces news primarily from policy seminars and briefings in capital city is a major impediment to communicate climate change in Bangladesh through media.

Journalists' knowledge

Extent and quality of coverage of any issue largely depend on "journalists' and editors' assumptions about what interests their audiences and what audiences feel affects their own interests" (O'Loughlin & Hoskins 2010:43). News media people make the assumptions on the basis of some so-called news values which are not universal and can vary between different journalistic cultures. Thus, what makes news and what's not is a major debate in journalism studies for long and the debate is going on without any concrete conclusion. However, one factor is important that shape the coverage – journalists' knowledge about the subject they cover. This is particularly true if the issue is a specialised subject area such as science. Content analysis of Bangladeshi newspapers reveals that most of the news reports are based on daily events (86.4 per cent) while proportion of investigative reports is only 3.6 per cent (Mahmud 2010). Lack of journalistic expertise to investigate the issue is the main reason behind limited in-depth coverage. More precisely, the reasons are journalistic lack of knowledge and training to understand the complexity of climate change science which is coupled with limited resources of the media organisations resulting in dependency on event-based news reports. The problem in fact lies within the media system of Bangladesh.

Media as an institution is still in its infancy in Bangladesh as the industry suffers from financial viability, inadequate audiences and readers, and lack of skilled human resources. Bangladeshi journalistic culture is dominated by politics which is rooted in the country's struggle against political causes of British colonial rulers, democratic and independence movements during the Pakistani era, and political freedom against military dictators until 1990 (Shoesmith & Mahmud 2009). Not surprisingly, politics is the main interest for media while environment in general and climate change in particular received little attention from the news policy makers until recent times. There is hardly any reporter who is specifically assigned in Bangladeshi news media outlets to cover science, although there is a recent trend of creating environmental beats in many newspaper and television news rooms. In most cases, one journalist is assigned to oversee all the events happening around environmental issues (including climate change). And at the worst, the environmental journalist is often assigned to cover general news events. In such a situation, Bangladeshi journalists have failed to acquire expertise about the complexities of climate change science as well as socio-economic and ecological impacts of changing climate.

The situation is much disappointing among the journalists who work outside the main newsroom of the capital or in the rural areas. They are known as *mofussil* journalists and send news from different parts of the country working as local correspondents or freelancers. Most of them are based in district levels and assigned to cover all newsworthy events of their locality, albeit their focus is always in politics and crimes of respective areas. In the Bangladeshi context, its coastal areas are most vulnerable to any real or possible consequences of climate change and we have already mentioned about little coverage of climate change from these hotspots. We argue that *mofussil* journalists' low levels of knowledge about climate change, work routines and limited training in journalism in general are the main reasons for the limited coverage.

Conclusion and Future Directions

In this chapter, we attempted to initiate a debate to study theoretical concepts of climate change communication as a dual field of risk and science communication. A merger of these two important streams of communication science will definitely contribute to develop better theoretical knowledge of climate change media communication. Here, we only instigated the debate by presenting some of the key features of risk and science communication that, we acknowledge, are subjects of further scrutiny.

Second, which was actually core of the study, we examined some of the major challenges to climate change communication through mass media in Bangladesh. Climate change, no doubt, is a communication challenge both for advanced and developing countries. But, as we have argued throughout the chapter, the extent of challenges multiplies in developing countries like in Bangladesh once socio-economic conditions and journalistic hazards are added with the challenges inherited from the climate change issue itself. The task of translating or providing better knowledge on global warming and climate change issues in Bangladesh gets even more difficult for the media given low literacy rate as the barrier. Even if we take broadcast media as tools to disseminate news to the public, it is often difficult for the people to understand meaning of those scientific jargons related to climate change.

If western media are accused of creating the debate that climate science is uncertain, media in Bangladesh clearly situated on the other side of the coin where climate change is 'absolutely certain' and 'already badly hit Bangladesh'. We believe both 'understatement of science' and 'exaggeration of science', what Storch (2009) rightly says cultural construct of climate change science, are major communication problems for public to understand. Media in Bangladesh, taking the radical position in line with its policymakers and activists, portrayed only the exaggerated episode of climate science by frequently relating natural hazards as results of global warming which has yet to have any credible scientific basis. Bangladeshi media facilitated such politicisation of climate change by representing government policymakers and environmental activists extensively as key definers of climate science where both scientists and lay people have failed to get any space to raise their voices. Furthermore, it has been found that limited journalistic capacity and dependency on Western media sources are major impediments to effective climate change coverage. Frequent recycling of articles on climate change issues from European and US media shape the global aspects of the issue undermining local viewpoints.

As a remedy, media organisations in Bangladesh can form networks with scientific communities, policy forums and activists (e.g., environmental NGOs) to hold dialogues and in-profession training of journalists. As argued by McCright and Shwom (2010), scientists will learn journalistic norms and journalists will learn scientific norms through such interactions. And, once the journalists acquire knowledge of the issue, they will find more perspectives to write about it. Harbinson and colleagues (2006) argued in the similar way as their study on climate change reporting in developing countries revealed that lack of journalistic training for specialised environmental issues decreased the number of climate change stories in developing countries' news media. The news organisations can also assign its journalists in specialised science beat and organise spot visits to vulnerable areas together with scientists. Any such initiative to enhance journalistic expertise should focus on *mofussil* journalists who report from rural areas. As a long term strategy, we also propose to introduce climate change communication in the curricula of journalism schools in Bangladesh through which the future journalists will get a solid foundation of the issues, complexities and local aspects of climate change impacts. Thus, increased journalistic knowledge and institutional capacity of media will pave the way for more climate change stories from local perspective involving different stakeholders that include scientists and lay people. Such mediated interaction of scientists, policymakers and lay people would allow all stakeholders' engagement in the discursive construction of climate change knowledge.

Finally, this is expected that this study will encourage further research on climate change communication in Bangladesh and other developing countries to understand media's role in constructing public perception, attitude and behavioural actions towards adaptation of climate change. Any such study should also consider influence of interpersonal communication (e.g., information coming from social contacts and different agencies – NGOs, government etc.) and personal exposure to hazards (e.g., living in a vulnerable area or experience of natural disasters) in the overall process of public perception and responses to climate change.

Acknowledgement

The authors would like to thank the Cluster of Excellence "Integrated Climate System Analysis and Prediction" (CliSAP) of the University of Hamburg for providing financial support for the research presented in this chapter.

References

Ali, A. 1999. Climate change impacts and adaptation assessment in Bangladesh. *Clim Res,* 12, 109–116.
Antilla, L. 2005. Climate of scepticism: US newspaper coverage of the science of

climate change. *Global Environmental Change*, 15, 338–352.

Bauer, M. 2009. The evolution of public understanding of science – discourse and comparative evidence. *Science Technology Society,* 14, 221–240.

BCCSAP 2009. *Bangladesh Climate Change Strategy and Action Plan*. Ministry of Environment and Forest, Government of Bangladesh (GoB), Dhaka. Available at: www.moef.gov.bd/climate_change_strategy2009.pdf [accessed: 10 May 2011).

Bell, A. 1994. Media (mis)communication on the science of climate change. *Public Understanding of Science,* 3, 259 – 275.

Bernstein, L. Bosch, P. Canziani, O. et al. 2007. *Climate change 2007: Synthesis Report,* IPCC: Geneva.

Billet, S. 2010. Dividing climate change: global warming in the Indian mass media. *Public Understanding of Science*, 99 (1–2), 1–16.

Bord, R. J., Fisher, A. & O'Connor, R. E. 1998. Public perceptions of global warming: United States and international perspectives. *Climate Research,* 11,75–84.

Boykoff, M. 2007. Flogging a dead norm? Newspaper coverage of anthropogenic climate change in the United States and United Kingdom from 2003 to 2006. *Area*, 39(4), 470–481. 2010. Carbonundrums: The Role of the Media, in *Climate Change Science and Policy,* edited by S. Schneider, A. Rosencranz, M. Mastrandrea and K. Kutz-Duriseti. Island Press, Washington, DC: Island Press, 397–404.

Boykoff, M. and Boykoff, J. 2004. Balance as bias: global warming and the US prestige press. *Global Environmental Change*, 14, 125–136. 2007. Climate change and journalistic norms: A case-study of US mass-media coverage. *Geoforum*, 38(6), 1190–1204.

Boykoff, M. and Mansfield, M. 2008. Ye Olde Hot Aire: reporting on human contributions to climate change in the UK Tabloid Press. *Environmental Research Letters* 3 (8). 2010. *2004-2010 World Newspaper Coverage of Climate Change or Global Warming*. Boulder, CO, University of Colorado. Available at http://sciencepolicy.colorado.edu/media_coverage/ [accessed: 20 March 2011].

Boykoff, M. and Roberts, J. 2007. Media coverage of climate change: current trends, strengths, weaknesses. *Human Development Report 2007/2008 – Fighting climate change: Human solidarity in a divided world, 2007/3*, 1–53.

Briscoe, M. 2004. *Communicating uncertainty in the science of climate change: an overview of efforts to reduce miscommunication between research community and policymakers & public*. International Centre for Technology Assessment, Washington.

Carvalho, A. 2005. Representing the politics of the greenhouse effect; discursive strategies in the British Media. *Crit Discourse Stud.* 2(1), 1–29. 2008. *Communicating Climate Change: Discourses, Mediations and Perceptions*. Braga: Centro de Estudos de Comunicação e Sociedade, Universidade do

Minho. Available at: http://www.lasics.uminho.pt/ojs/index.php/climate_ change [accessed: 13 February 2011].

Carvalho, A. and Burgess, J. 2005. Cultural circuits of climate change in U.K. broadsheet newspapers, 1985–2003 . *Risk Analysis*, 25(6), 1457–1469.

Dirikx, A. and Gelders, D. 2010. To frame is to explain: A deductive frame-analysis of Dutch and French climate change coverage during the annual UN conferences of the parties. *Public Understanding of Science*. Available at: doi:10.1177/096366250935204 [access: 23 March 2010].

Dispensa, J. and Brulle, R. 2003. Media's social construction of environmental issues: focus on global warming. *International Journal of Sociology and Social Policy*, 23(10), 74–105.

Downs, A. 1972. Up and down with ecology – the "issue-attention cycle". *The Public Interest*, 28(Summer), 38–51.

Filho, W. 2008. Communicating climate change: challenges ahead and action needed. *International Journal of Climate Change Strategies and Management*, 1(1), 6–18

Fischhoff, B. 1995. Risk Perception and Communication Unplugged: Twenty Years of Process. *Risk Analysis,* 15(2),1471–82.

Giddens, A. 2008. The politics of climate change: National responses to the challenge of global warming. *Policy Network Paper*. Policy network: UK. 2009. *The politics of climate change*. Cambridge: Polity Press.

Gurabardhi, Z. Gutteling, J. and Kuttschreuter, M. 2004. The development of risk communication: an empirical analysis of the literature in the field. *Science Communication,* 25, 323–349.

Haldén, P. 2007. *The geopolitics of climate change. Challenges to the international system.* FOI, Swedish Defence Research Agency: Stockholm.

Hansen, A. 1994. Journalistic practices and science reporting in the British press. *Public Understanding of Science*, 3(2), 111–134.

Harbinson et al. 2006. *Whatever the weather: media attitudes to reporting climate change*. London, UK: Panos Institute. Available at: http://www.panos.org. uk/?lid=308 [accessed: 20 April 2011].

Houghton, J. 2004. *Global Warming: The Complete Briefing*. Cambridge: Cambridge University Press.

Irwin, A. and Wynne, B. 1996. Introduction, *Misunderstanding science? The public reconstruction of science and technology,* edited by A. Irwin and B. Wynne. Cambridge: Cambridge University Press, 1–17.

Kempton, W. 1991. Public understanding of global warming. *Society and Natural Resources,* 4(4), 331–345.

Kitzinger, J. 1999. Researching risk and the media. *Health, Risk and Society,* 1(1), 55–69.

Koivurova, T. Keskitalo, E. C. and Bankes, N. 2009. *Climate governance in the Arctic*. Dordrecht: Springer.

Kuha, M. 2009. Uncertainty about causes and effects of global warming in U.S. news coverage before and after Bali. *Language & Ecology*, 2(4).

Leiserowitz, A. 2006. Climate Change Risk Perception and Policy Preferences: The Role of Affect, Imagery, and Values. *Climatic Change,* 77 (1–2), 45–72.

Leiss, W. 1996. Three phases in the evolution of risk communication practice. *The ANNALS of the American Academy of Political and Social Science,* 545, 85–94.

Leshner, A. 2003. Public engagement with science. *Science,* 299, 977.

Lorenzoni, I. and Pidgeon, N. 2006. Public views on climate change: European and USA perspectives. *Climatic Change,* 77 (1), 73–95

Mahmud, S. (2010). *Climate change coverage in Bangladeshi newspapers – a content analysis.* Master Thesis submitted to the Institute of Journalism and Communication Studies, University of Hamburg.

Mazur, A. and Lee J. 1993. Sounding the global alarm: environmental issues in the US national news. *Social Studies of Science,* 23(4), 681–720.

McComas, K. and Shanahan, J. 1999. Telling stories about global climate change: measuring the impact of narratives on issue cycles . *Communication Research,* 26(1), 30–57.

McCright, A. M. and Dunlap, R.E. 2003. Defeating Kyoto: The Conservative Movement's Impact on U.S. Climate Change Policy. *Social Problems* 50(3), 348–373. 2010. Anti-reflexivity: the American conservative movement's success in undermining climate science and policy. *Theory, Culture & Society,* 27, 100–133. 2011. The politicization of climate change and polarization in the American public views of global warming, 2001-2010. *Sociological Quarterly,* 52 (2), 155–194.

McCright, A. and Shwom, R. 2010. Newspaper and television coverage. in *Climate Change Science and Policy,* edited by S. Schneider, A. Rosencranz, M. Mastrandrea and K. Kutz-Duriseti. Island Press, Washington, DC: Island Press, 405–413.

McManus, P. 2000. Beyond Kyoto? Media representation of an environmental issue. *Australian Geographical Studies,* 38(3), 306–319.

Mirza, M. 2003. Climate change and extreme weather events: can developing countries adapt? *Climate Policy,* 3 (3), 233–248.

Nagel, J. Dietz, T. and Broadbent, J. 2008. *Sociological perspective of global climate Change.* Proceedings of the workshop: Sociological Perspectives on Global Climate Change, Virginia, 30–31 May 2008.

NAPCC 2009. *National Action Plan on Climate Change (India).* Prime Minister's Council on Climate Change, India, Prime Minister's Office. Available at: http://pmindia.nic.in/climate_change.htm. [accessed: 10 March 2011]

Neverla, I. 2007. The birth of European public sphere through European media reporting of risk communication. *European Societies,* 9(5), 705–718. 2008 *The IPCC reports 1990-2007 in the media: A case study on the dialectics between journalism and natural sciences.* Paper to the ICA conference: Global Communication and Social Change, Montreal, 22–26 May 2008.

Neverla, I. and Schäfer, M. (Hrsg.) 2012. *Das Medien-Klima: Fragen und Befunde der Kommunikationswissenschaftlichen Klimaforschung.* Wiesbaden: Verlag für Sozialwissenschaften.

Olausson, U. 2009. Global warming – global responsibility? Media frames of collective and scientific certainty. *Public Understanding of Science*, 18, 421–436.

Renn, O. 1991. Risk communication and the social amplification of risk, in Communicating risks to the public: international perspectives, edited by Kasperson, R. and Stallen, P. Netherlands: Kluwer Academic Publishers.

Schäfer, M. Ivanova, A. and Schmidt, A. 2011. Global Climate Change, Global Public Sphere? Media Attention for Climate Change in 23 Countries. Studies in Communication Media (SCM), 0. Jg., 1/2011, 131–148.

Shanahan, M. 2009. Time to adapt? Media coverage of climate change in nonindustrialised countries, in Climate change and the media, edited by Boyce, T. and Lewis, J. New York: Peter Lang. 145–157.

Shoesmith, B. & Mahmud, S. 2009. Bangladeshi Mediascape. In: Hans-Bredow-Institut (ed.): *Internationales Handbuch Medien.* [International Media Handbook] Baden-Baden: Nomos, 810 – 824.

Smith, J. 2005. Dangerous news: media decision making about climate change risk. Risk Analysis, 25 (6), 1471–1482.

Stamm, K. Clark, F. and Eblacas, P. 2000. Mass communication and public understanding of environmental problems: The case of global warming. Public Understanding of Science, 9, 219–37.

Storch, H.v., 2009. Climate research and policy advice: scientific and cultural constructions of knowledge. Env. Science Pol. 12, 741–747

Sturgis, P. and Allum, N. 2004. Science in society: Re-evaluating the deficit model of public attitudes. Public Understanding of Science, 13, 55–74.

Trumbo, C. 1996. Constructing climate change: claims and frames in US news coverage of an environmental issue. Public Understanding of Science, 5, 269–283.

UNDP 2004. A Global Report: Reducing Disaster Risk – A Challenge for Development. Available at: http://www.undp.org/cpr/disred/rdr.htm [accessed: 14 April 2010].

Ungar, S. 1992. The rise and (relative) decline of global warming as a social problem. The Sociological Quarterly, 33(4), 483–501.

Ward, B. 2003. Reporting on climate change: understanding the science.

Washington, D.C.: Environmental Law Institute.

Webster, M. 2002. The curious role of "learning" in climate policy: should we wait for more data? Energy Journal, 23(2), 97–119.

Weingart, P. Engels, A. and Pensegrau, P. 2000. Risks of communication: discourses on climate change in science, politics, and the mass media. Public Understanding of Science, 9, 261–283.

Wilson, K. 2000. Communicating climate change through the media: Predictions, politics, and perceptions of risks, in Environmental Risks and the Media, edited by Allan, S. Adam, B. and Carter, C. New York: Routledge. 201–217.

World Bank 2009. Bangladesh at a glance – World Bank Group. Available at : devdata.worldbank.org/AAG/bgd_aag.pdf [accessed: 5 April 2011].

Chapter 4.2

Could Films Help to Save the World from Climate Change? A Discourse Exploration of Two Climate Change Documentary Films and an Analysis of their Impact on the UK Printed Media

Gabriela Ramirez Galindo

Introduction

The climate change debate has moved out of the scientific field and reached a pinnacle point with awareness raisers, activists and other supporters as well as sceptics voicing opinions at all levels of society, among which the most prominent are politicians and media.

In the cinema field, the documentation of the possible environmental and socioeconomic challenges that humans and nature will have to face as a consequence of climate change seems to attract the usual large producers like BBC or Discovery, independent producers, and even other non traditional film producers including political leaders like Al Gore and Prince Charles, film stars like Leonardo DiCaprio, singers like Björk and eco-activist groups like Greenpeace.

The recent proliferation of climate change films coincides with the growing scientific consensus that there is an important human-induced factor which hastens climate change, yet some climate change denial documentaries have also been made. Acknowledging the production of many other visual formats of climate change movies, this chapter focuses on two big screen long format documentaries, created for mainstream circuits, aimed to impact on global audiences.

One is a top-down initiative led by the internationally well known politician Al Gore, whose documentary film, *An Inconvenient Truth* (2006), is one of the earliest, best-known and most cited films about climate change. The other is a bottom-up initiative coming from an individual independent director, Franny Armstrong, whose documentary film *The Age of Stupid* (2009) relied on a 'crowd funded scheme' [1] of 242 investors (The Age of Stupid Net 2009: no page no.).

---

1   An investment scheme where funders get a percentage of the profits once the film makes net profits. Detailed information available at http://www.ageofstupid.net/node/1253/

This chapter analyses the portrayal of climate change risks through cinema, which I propose to denominate 'climate change-ism' discourse. The objective is to contribute critically to the understanding of the representation of climate change and how it is leading to the formation of a particular discourse which echoes and resonates indirectly in other media. The overall proposal is relevant in an undoubtedly ongoing climate change debate and in the search for ways of addressing social, economic and environmental problems envisaged in a not so distant future.

Theoretical Part

The theoretical section consists of four main parts aimed at understanding the formation of a discourse in films that support the climate change thesis. Mediation, representation and their implications are reviewed in relation to climate change. The theoretical examination is assisted by Beck's (1995) concept of 'risk society', which offers the basis for the formation of a discourse characterized by the anticipation of the climate change risk becoming reality and the creation of a cosmopolitan solidarity based on a global common fear of global destruction. To understand the implications of environmental discourses, the relationship between media and the public opinion on climate change is reviewed. Towards the end, this chapter offers an exploration of the few existing analyses about eco-environmental films, which gives an indication of the insufficient discussion in this respect, and proposes a framework approach to climate change discourse in films.

Climate Change Mediation and Representation

In recent decades, nature and environmental issues have failed to continuously maintain high public and media attention even when they have become a constant field of public and political debate and concern (Cottle 1993, Hansen 1993, Anderson 1997, Smith 2000). For example, by the beginnings of the 1990 decade, Hansen (1993) observed that the higher or lower levels of public and political concern were not directly related to the state of environmental degradation.

Media impacts on our understanding of the nature and environmental issues, however it is inseparable from other contextual, political, social and economic factors, like social and power structures or the communication activities of sponsors namely politicians, scientists or environmentalists (Hansen 1993, Anderson 1997, Smith 2000).

The mediation process inevitably requires the presence of media organizations (Thompson 1995), the exercising of and resistance to power structures (Silverstone, 2005) and a broader sociocultural context (Lindahl Elliot 2006). According to Macdonald (2003) our meaning making process of words and images is determined by a broader sociocultural context. This implies building connections between the reality, our mental concepts and the language we use to refer to words and images

(Hall 1997), and leads to the creation of 'discourses'. According to Hall (1992), a discourse is 'a group of statements which provide a language for talking about — i.e. a way of representing— a particular kind of knowledge about a topic' (291). This discourse enables us to construct the topic in a certain way, and limits the other ways in which the topic can be constructed (Hall 1992).

Discourses as systems of representation, construct an oversimplified, divided and fragmented world, while the reality embraces complex and innumerable historical, cultural, political and economic factors. However, discourses produce knowledge, influence social practices and are always embedded in the organization of power relations, which means discourses are part of power being exercised and power being resisted (Hall 1992: 295).

For Foucault (1972), discourses can be produced by different individuals in different contexts, always necessarily implying the taking of a position. For him, the group of statements that form discourses are systems, working in regular and systematic relationships and differences, and in constant change and transformation where old meanings continue interacting with new ones.

Climate change discourses, for example, instead of just being an homogeneous discourse about scientific facts, it has become a debate of moral, political and power positioning about believing it or not among media, politicians and public opinion. While climate change theory defenders' discourse argues for a worldwide responsibility and for the need of immediate global action before we arrive at an irreversible tipping point, detractors' discourse states that climate change still lacks scientific proof, with some detractors claiming that such global action has an 'unintended consequence of prolonging endemic poverty and disease in the Third World' (Channel 4 2009).

Lindahl Elliot (2006) dates the rise of a 'mediated environmentalism' to the 1960s when Rachel Carson published *Silent Spring* in 1962, a book about pesticides and its effects on nature and humans, that caused reactions of the chemical industry while contributing to a public debate. According to Wilson (2000), the portrayal of global warming news started with catastrophism and dramatic overstatements almost as an accident resulting from the combination of a declaration done by the scientist Jane Hansen in June 1988 affirming that global warming was here and a hot summer ended in a drought (205).

Images of threatened environments symbolizing the hazards to life have been common during the last 30 years as a response to risks like global warming (Allan et al. 2000), and for Hansen (1993), the mediation and representation of the environment were Western oriented and defined in terms of 'anything nuclear', pollution or conservation of endangered species until the beginning of the 1990s, when the environmental debate started moving towards a global consciousness perception with global warming highlighting the new news agenda (xvi).

The 'Risk Society' Concept, Climate Change as a Democratic Risk

'Climate change exacerbates existing inequalities ... but simultaneously dissolves them' (Beck 2009: 11).

For Beck, the creator of the 'risk society' concept, every culture, ethnic group, religion and region in the world faces for the first time in history a future that threatens one and all. This has consequences in the political realm, challenging the logic of the nation-state and the industrial society, because individuals and nations can no longer constrain themselves thinking in terms of geographic limits and distance. Therefore, the politics of climate change are necessarily inclusive and global (Beck 2009).

For Beck (2009) the mediation of risks comes in the form of a '*Cosmopolitan event*', a spontaneous addressing of catastrophes, in real time and at global scale, where the traumatic experience overcomes the geographical distances and creates a cosmopolitan solidarity (e.g. a tsunami), or as an '*Enforced cross-border communication*', a unification forced by threats as condition rather than choice, a 'negative solidarity based on the fear of global destruction' (11–12).

Climate Change and Public Opinion

There is a general agreement that media plays an influencing role on environmental public opinion. For some researchers this relation is linear-causal (see Parlour and Schatwow 1978, Brosius and Keplinger 1990, Nisbet and Myers 2007), while for others there is a lack of conclusive evidence about the effects over environmental public opinion and attitudes (Anderson 1997).

For Nisbet and Myers (2007), who reviewed the trends in public opinion about global warming in the USA during 20 years, 'there are consistent connections between patterns of media attention to global warming and shifts in poll trends' (445). But for researchers like Anderson (1997), such conclusions are the result of a linear relation analysis in identifiable stages rather than interactive studies (178). The fact that most of the research has been done in industrialized countries (Anderson 1997) poses some questions to the validity of the findings when applied to developing contexts, where education is not generalized and media penetration and professionalization is uneven.

Other researchers affirm that despite the efforts the evidence is still insufficient. Shanahan (1993), for example, investigated the relation between television exposure and the level of environmental concern in USA between 1988 and 1992. His conclusions suggested that even when mass media could play a key role in how environmental messages are received and interpreted, television exposure does not have an impact on the raising of environmental concern (195). Other studies by Gunter and Wober (1983), Wober and Gunter (1985) and Protess et al. (1987) suggest similar conclusions.

The spiral of silence model proposed by Noelle-Newman (1974, 1993) is significant for considering simultaneously the influence of mass media, interpersonal and social relations, individual expression and individual perception of the dominant surrounding opinion. However, it has received criticism for being considered with a limited scope that 'tends to gloss over the complexities of the processes by which different social groupings make sense of environmental meanings as presented in the media' (Anderson 1997: 178). For Anderson (1997), some of those groups can also play a potentially influential role over media and political agendas beyond receiving and appropriating environmental meanings from the media (See Hansen 1993 analysis on the coverage given to Greenpeace between 1987 and 1991 in two British National daily papers).

The widespread use of mass media by environmental pressure groups correspond to the emergence of the 'modern environmental concern' in the 1960s, characterized by scientists widely spreading discoveries like the hole of the ozone layer and its environmental consequences, and a public opinion reliant on the mediation and representation or environmental issues offered by a variety of sources (Hansen 1993: 150).

Climate Change and Environmental Cinema

Ingram (2000) undertook one the few studies available on environment and cinema. He analysed more than 150 Hollywood movies with an evident or central environmental topic and identified three central themes ('Wilderness', 'Wild Animals' and 'Development and the Politics of Land') that reflect how environmental American ideology has perpetuated romantic attitudes towards nature and played a role in the circulation of environmental discourses.

More recently, Murray and Heumann (2009) reflected on what they call 'contemporary popular environmentalist films', formed by films that provide obvious ecological messages contained in comedy or melodrama genres. For them, films like *The Day After Tomorrow* (2004), *Happy Feet* (2006), and *Ice Age: The Meltdown* (2006) are the representatives of a new wave of films characterized by recurrent themes like environmental disasters and a multiplicity of environmental risks.

The lack of "climate change and cinema" references in the literature, shows that probably because of its relative novelty, the general characteristics of a climate change discourse in films, what I propose to call 'climate change-ism', is a topic with potential for future research.

The Research Questions

This chapter works with the assumption that climate change films are developing a particular discourse in order to contribute to the formation of public opinion. To operationalize this hypothesis two contrasting long format (90 minutes) documentaries, *An Inconvenient Truth* (2006) and *The Age of Stupid* (2009) were analyzed. While both films aim to reach large audiences and raise awareness about climate change and its consequences, as well as to promote environmentally friendly behaviours, they clearly use distinct approaches. *An Inconvenient Truth* is presented in a top-down style by the former Vice President of USA, Al Gore; *The Age of Stupid* presents a bottom-up approach where the life of six individuals is documented and observed in 2055 by an imaginary man looking back at real footage taken in the present.

Based on the main question 'How do films contribute to the formation of public opinion in relation to climate change?' two subquestions were explored: 1) What are the distinct and common characteristics of a climate change discourse in both films in relation to power and social change? and 2) Do the films as information sources for other media contribute differently towards the formation of public opinion on climate change?

Research Design and Methodology

To investigate the two research questions, two complementary analyses were carried out. A Critical Visual Discourse Analysis was used to analyze sequences in each film. Visual Discourse Analysis as a research methodology focuses its attention on images themselves and the way they interact with other elements like text. Its central concern is the production of discourses and concentrates on questions of power and the way images and texts are articulated (Rose 2001).

To understand to which extent both movies became sources of information for other media and whether their contribution was in relation to the discourse exposed in the films, articles from the four main UK broadsheets were analyzed with the methodology of Content Analysis. Content Analysis, as a systematic and quantitative method, allows obtaining a general overview of frequencies and prioritization or patterns of framing the information (Hansen 1998, Deacon 1999).

Operationalizing the Research Questions, Climate Change-ism

To operationalize the subquestions this chapter investigated four central aspects considered, after thorough viewing, as the focal points of insights for the understanding of a climate change discourse, here denominated climate change-ism:

a. *Credibility.* Who are the spokespeople selected to give the climate change message?
b. *Portrayal of climate change.* How are the risks of climate change portrayed?
c. *Addressing resistance to change.* How do they portray the resistance to chance?
d. *Urgency of action.* Do they deliver a call to action message? How is it constructed?

The Selection of Data and Design of Research Tools

As visual analysis of the entire films would be an extremely time consuming strategy, sequence (as proposed by Iedema 2001) was selected as the study unit. A sequence consists of a camera moving 'with specific character(s) or subtopic across time-spaces' (189). As an intermediate unit, it allows discussing the unifying characteristics of each film while permitting analysis of detailed elements in their context. The analysis followed the critical analysis of media discourse framework proposed by Fairclough (1995).

Originally eight sequences were identified from each film. Only four were selected and analysed because they were comparable sequences in both films. Each pair of sequences corresponded to one of the four elements of climate change-ism discourse proposed.The final sequences allowed contrasting the approach of the films (top-down vs. bottom-up) while uncovering the parallelisms on the topics they aim to communicate to their audiences like urgency of action.

For the content analysis the four main broadsheets of the UK were selected to measure the influence of the films at national level, The Guardian, The Times, The Independent and The Daily Telegraph. The period of analyses included one month before and two after the release of the films in the UK, 15 September 2006 for *An Inconvenient Truth*, and 15 March 2009 for *The Age of Stupid*. Advertising, reply letters, repeated publication of the same articles and non analyzable single sentence references were eliminated from the samples. *The Age of Stupid* sample (Sample 1) consisted of 31 articles complying with the criteria, from 39 originally obtained. The *An Inconvenient Truth* sample (Sample 2) contained 40 articles, one out of every two articles complying with the criteria, from an original total of 142 articles. The variables were either nominal or ordinal and 42 per cent of the articles were analysed by an external coder. The coders achieved a score of 89 per cent in the inter-coder reliability test.

Results and Interpretation

1. Critical Visual Discourse Analysis

In both films, climate change risks find expression in four identified common characteristics of a discourse here denominated climate change-ism. They can be regarded as progressive steps that start with a) advice coming from authority or credible voices, through the portrayal of b) apocalyptic visions and c) the risks of denial, ending in d) an urgent salvation message (Figure 4.2.1).

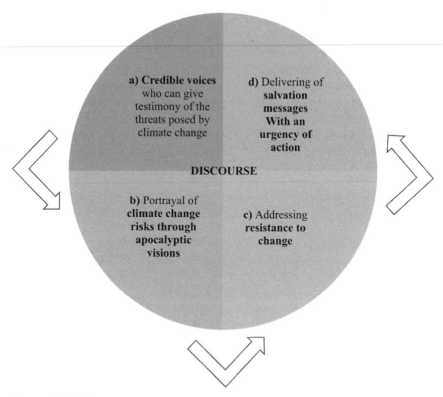

Figure 4.2.1 Four connected aspects of Climate change-ism

a) Selection of Credible Voices as Spokespeople who Can Give Testimony of the Threats Posed by Climate Change

The approaches used by the films were contrasting (top-down vs. bottom up), therefore their testimonies of climate change threats rely on different constructions of credible spokespeople. In *An Inconvenient Truth*, the authority voice is Al Gore, whose discourse is composed by a combination of political, scientific and emotional arguments, using evidence from scientific contacts and political visions of the world. By contrast *The Age of Stupid* gives voice to normal people living in developed and developing regions (the UK, The Alps, Nigeria and Iraq) through first hand credible testimonies of the damages posed and suffered due to climate change, aided by an external voice coming from the future who politically and ideologically frames the testimonies presented.

In the chosen sequence from *The Age of Stupid*, Layefa, one of six main characters, represents marginalized groups and the victims of capitalism, corruption and environmentally unsustainable human behaviour. She is a young black girl in the poor African country of Nigeria. Layefa is constructed as a credible and reliable person when she actively explains to the camera her village daily life '*The fishing is not good, because of the oil spills there killing most of the fish'.* Her testimony of suffering and struggles in an oil region is framed and extended by a narrator, who engages the audience in the criticism of the oil industry in which Shell corporation, whose logo appears three times in the sequence, becomes a symbol of capitalism and a consistent system of exploitation of people and resources in developing countries while contributing to climate change.

In *An Inconvenient Truth*, after being introduced as a star to the audience, Gore positions himself by saying: '*I used to be the next president of the USA*', which locates him as a high profile, internationally influential politician, but also implies a lamenting of his own political defeat.

The politician also constructs himself as a 'charismatic and sympathetic character', and as the aspiring messiah of climate change, '*I've been trying to tell this story for a long time and I feel as I've failed to get the message across*'. What the audience witnesses is a contradictory character, a doubly frustrated man, defeated in a presidential position, and defeated in his moral and personal political battle against climate change deniers, yet at the same time being an influential well positioned politician (who later shared the Nobel Prize 2007 for this movie) on which he builds his authority (top-down).

b) Portrayal of Climate Change Risks through Apocalyptic Visions

In order to deal with the possibilities of being confronted by the environmental self induced destruction, both films use Hurricane Katrina as a didactical and emotional example showing that also powerful nations like USA, should be worried about the magnitude and intensity of the consequences of climate change.

In *The Age of Stupid,* Hurricane Katrina hitting New Orleans is framed with rapid shots and an American flag waved by the potent tempest followed by blown down street name signs of two American symbols, Martin Luther King and Hollywood. The disaster depiction is mostly based on the moving and emotional first hand testimony of Alvin Duvernay, a bottom-up hero who explains his experience on his boat at tree level in a swamped neighbourhood full of people waiting for help after the hurricane.

In *An Inconvenient Truth,* followed by a rational explanation about the connection between ocean temperatures and storms, Gore becomes emotional and nationalist political. The damage done to the American economy is prioritized by showing the damage experienced by the oil industry. Then, after shots of the disaster, we see Gore with his back to the camera asking sadly '*How in God's name could that happen here?*' Distant sufferers and remote threats are no longer far away, Americans themselves become sufferers and vulnerable to unmanageable hazards.

Despite contrasting narratives, in both films Hurricane Katrina is used to help the audiences to visualize a possible future in which large scale environmental events can alter the political and economic life of nations, transform entire cities and affect millions of human lives. The messages can be compared to Becks notion of 'cosmopolitics', where the greater the planetary threat is, the greater the possibility or reaching even the richest, the wealthiest and powerful countries.

c) Resistance to Change

In *An Inconvenient Truth* the references are mainly political. Gore constructs himself as a politician embarked on a continuous and tiring effort of travelling the world with a salvation message for stakeholders. To emphasize the fatigue of repeating endlessly the same message and to affirm that the world seems to be stubbornly deaf to his message, the camera follows him in a routine through the airports.

Another resistance portrayed is the lack of political will. Gore positions himself as opposite to Ronald Reagan and George H.W. Bush. The camera makes a close up of Gore's reflective face and his working hands while in voiceover he affirms '...It's easy for them to ignore it...to say: Well, let's deal with that tomorrow'. The image and the stress on the words turn into a direct criticism of leaders neglecting their responsibility, positioning himself as politician of actions rather than words.

In *The Age of Stupid*, the resistance to change is portrayed at the local and media level, where the latter is depicted as a contributor to a mostly irrational and emotional public opinion debate. A media reporter and a photographer being recorded by the film camera are observed by the audience while feeding the public debate by giving coverage and voice to a tiny group of protesters against wind turbines. The story of the reporter and the photographer seems to be constructed in an unbalanced way, due to an evident lack of opposition voices interviewed. The sequence itself seems to do something similar than what it criticizes, constructing the group of opposition as the only ones who can be over-emotional on this issue.

The group of wind turbines opponents is depicted giving mostly senseless contradictory statements: 'The problem really is that this is one of the least windy sites in the country', immediately contradicted by a colleague saying about the protest balloon they released in the scene 'I hope it's not going to get too windy tomorrow and it wraps itself around the church'.

d) Delivering of Salvation Messages with an Urgency of Action

It is remarkable that in the final sequences both movies present themselves as opening to the audiences a privileged window to envisage the future. Gore places himself at a first glance as a common man 'There's nothing that unusual about what I'm doing...' while in fact he is placing himself as a climate change messiah 'I had the privilege to be shown it as a young man'. The privileged window is emphasized visually in the film as a shape of his figure illuminated by a spiral of light, preceded by a voice announcing his coming.

In *The Age of Stupid*, the window of salvation is offered from the 'future hell of regret' by a fictional character, a survivor, who offers the audience the opportunity to look back to their own present, like an echo of Charles Dickens's 'A Christmas Carol', to glimpse through a window where an apocalyptic future (borrowing

Gore's words from the other film) is *'...very clearly visible. See that? That is the future in which you are going to live your life'*.

To engage viewers with a sense of urgency both films invoke regret feelings about the last chance lost and a concern about future generations blaming us. In *The Age of Stupid*, the Archivist asks *'Why didn't we stop climate change when we had the chance?'*, and Fernand, a 82 French mountain guide, affirms *'I think everyone in the future will perhaps blame us for not thinking to protect the environment'*. In *An Inconvenient Truth* Gore asks in a similar way *'Future generations may well have occasion to ask themselves: What were our parents thinking? Why didn't they wake up when they had a chance?'*

Resembling Beck's idea of 'cosmopolitics' (2007), Fernand, while climbing a high mountain with his grandchildren states *'In the mountains you're roped together. The risk is the same for you as it is for me'*, as metaphor that not even the most powerful or experienced can escape, climate change threatens one and all, elderly and child, large and small countries, East and West.

Beck's (2007) concept of cosmopolitics as a hazard connecting rich and poor, powerful and powerless, goes one step further in these films, connecting human generations in past, present and future.

2. Content Analysis

The Content Analysis explored whether the analysed films contributed differently to the media content about climate change, whether their discourse resonated and if they became climate change references for printed media during the period of their release.

Initially, the films contributed differently to the media content on climate change through being covered in different proportions by the broadsheets. In both cases, the most extensive coverage was given in The Guardian and The Times (Figure 4.2.2). However, in 2006 *An Inconvenient Truth* received a coverage more evenly distributed among the four media sources, while in 2009 *The Age of Stupid* received less coverage from The Independent and The Daily Telegraph.

At different levels, the celebrity voices of the films determined to some extent their impact on the media analysed. For *An Inconvenient Truth,* Al Gore was the main reference, mentioned in 100 per cent of the articles, while the Director, Davis Guggenheim, was obscured by the politician and mentioned in only 20 per cent of the articles and always in conjunction with Gore (Figure 4.2.3).

Gore's dominant presence also seems to have attracted wider media coverage, with 32 per cent of mentions in political articles and 48 per cent in entertainment articles. However, Gore as presenter generated differentiated reactions, the balance among positive, descriptive and ambiguous positions about him was similar, respectively 35, 25 and 30 per cent (Figure 4.2.4). In the same way, the film received a controversial judgement, with 30 per cent of the mentions being either ambiguous or combined positive and negative comments.

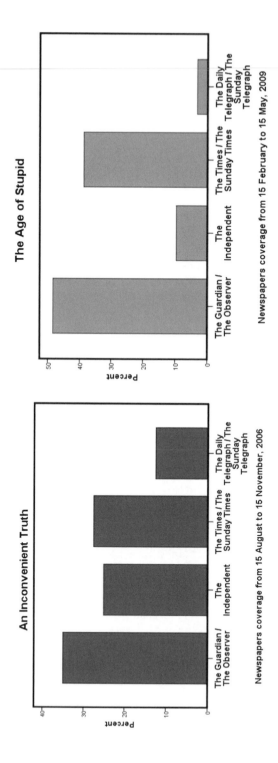

Figure 4.2.2 Coverage given to both films one month before and two after their release

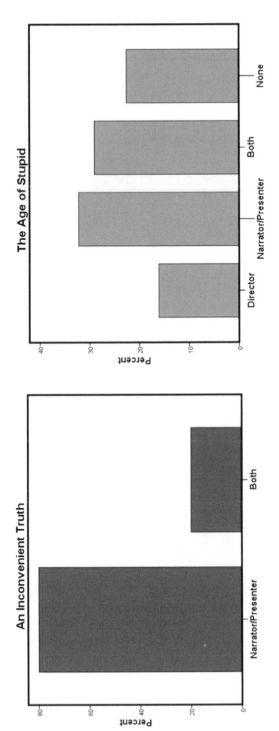

Figure 4.2.3 Mentions of Director and/or Narrator/Presenter from the films

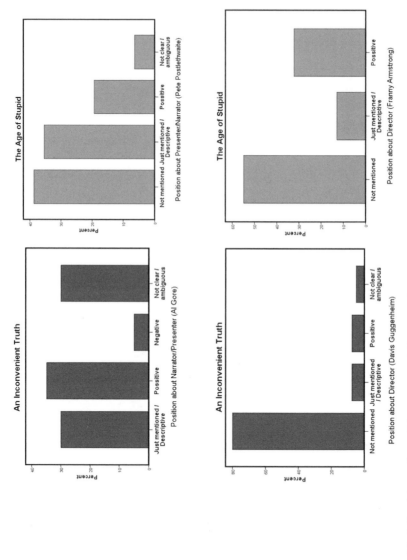

Figure 4.2.4 Positions about Directors and Narrator/Presenters

By 2009, the film seems to have achieved positioning itself as the main climate change film reference. In 2006, during its UK release, it was occasionally (10 per cent) compared with either other movies or eco-movies. Three years after, it seemed to be the common and logical reference when journalists compared the new climate change movie *The Age of Stupid* with previous films. *The Age of Stupid* was compared with other movies in 30 per cent of the articles mentioning it, and 88 per cent of those comparisons were with *An Inconvenient Truth.*

In *The Age of Stupid*, the narrator, Pete Postlethwaite, was the personality who most attracted the media attention. 32 per cent or the articles mentioned him alone, 16 per cent the Director alone, and 29 per cent both (Figure 4.2.3). Contrastingly, to the other film, *The Age of Stupid* obtained a coverage mainly centred in entertainment (68 per cent) and less mentions in political articles (26 per cent), with 80 per cent of the coverage being descriptive or positive. Neither of the films achieved to resonate significantly in articles with scientific or academic content (Figure 4.2.5).

One indication that the media accepted these films as climate change reference via their personalities is the fact that clear negative judgements were very rare with only Gore receiving 5 per cent of them. This acceptance of the films as climate change reference comes out even clearer in the value judgement towards the films, with both films receiving 45 per cent of clear positive articles and less than 8 per cent clear negative, a large difference.

Another indication that aspects of the core discourse of the films were considered relevant by the media is the fact that climate change was at least mentioned in the majority of the articles; 97.5 per cent for *An Inconvenient Truth,* and 84 per cent of the time for *The Age of Stupid.* On a first level this means that both films managed to get the issue of climate change on the agenda.

In 84 per cent of the articles that mentioned *The Age of Stupid* the climate change-ism discourse resonated, while in *An Inconvenient Truth* it was observed in only 61.5 per cent of the cases. Part of this difference could be attributed, three years later, to the growing mention of the climate change debate on news.

Conclusion

Climate change documentary films are part of a cinema trend whose aim, beyond contributing to the knowledge about the topic, is to promote environmental responsible attitudes and social change. The research model allowed to discover how the two films as media products contributed to the mediation and representation of the social understanding of climate change, and how echoing Beck's (1995, 2009) concept of 'risk societies', they visually constructed a discourse about the potential danger of parallel environmental and human destruction.

The environmental risks find expression in four identified common characteristics of a discourse here denominated climate change-ism. They can be regarded as progressive steps that start with a) an advice coming from authority or

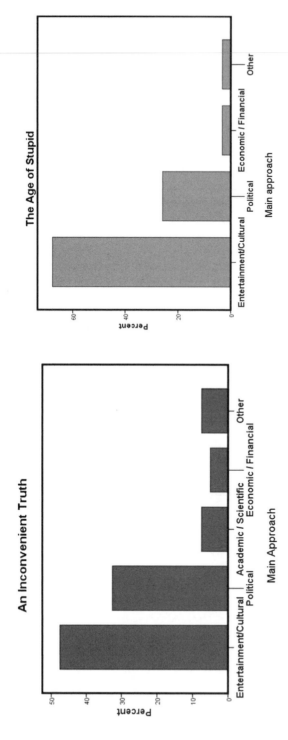

Figure 4.2.5 Main approach of the articles where references to the movies were included

credible voices, through b) the portrayal of apocalyptic visions and c) the risks of denial, ending in d) an urgent salvation message, which were confirmed as central elements in both films.

The approaches used by the films were contrasting (top-down vs. bottom up), therefore their testimonies of climate change threats rely on different constructions of credible spokespeople. In *An Inconvenient Truth*, the authority voice is Al Gore, whose discourse is composed of political, scientific and emotional fragments, using evidence from scientific contacts and political visions of the world. By contrast *The Age of Stupid* gives voice to normal people around the world through first hand credible testimonies of the damages posed and suffered due to climate change, aided by an external voice coming from the future who politically and ideologically frames the testimonies presented.

The content analysis allowed to appreciate that the films ensured climate change presence in the media, however, the articles were fairly equally divided between simple mentions and portraying the urgency of action conveyed in both films. The celebrity voices of the films determined to some extent their impact on other media, a factor particularly noticeable in *An Inconvenient Truth,* where the presence of Al Gore generated wider coverage in the entertainment and political spheres, and positioned the film as a political reference, however controversial. By 2009, the film seems to also have achieved positioning itself as the main climate change film reference.

To further explore the climate change-ism discourse in future research, discourse analysis could be used to a) compare other climate change films, and b) critically differentiate it from other traditional visual environmental discourses like those proposed by Ingram (2000) or Murray and Heumann (2009). Interviews to producers and audiences could offer the opportunity to uncover the power dimension behind the production of climate change films, explore how these and other films permeate in public opinion and social behaviours, and to analyse if the films are reaching their social change objectives. A follow up content analysis comparing longer time periods could give further insights of how the movies' discourse permeate into other media and how they become political and environmental references.

References

Allan, S., Adam, B. and Carter, C. (eds.) 2000. *Environmental Risks and the Media.* London: Routledge.

An Inconvenient Truth (dir. David Guggenheim, 2006).

Anderson, A. 1997. *Media, Culture and the Environment.* London: UCL Press.

Beck, U. 1995. *Ecological Politics in an Age of Risk.* Translated by Weisz, A. Cambridge: Polity Press.

Beck, U. 2007. In the new, anxious world, leaders must learn to think beyond borders. *The Guardian*, 13 July, section Comment and debate, 31.

Beck, U. 2009. Critical theory of world risk society: a cosmopolitan vision. *Constellations,* 16 (1), 3–22.

Brosius, H.B. and Keplinger, H.M. 1990. The agenda-setting function of television news: static and dynamic views. *Communication Research,* 17 (2), 183–211.

Carson, R.L. 1962. *Silent Spring.* New York: Fawcett Crest.

Channel 4. 2009. *The Great Global Warming Swindle.* [Online]. Available at www.channel4.com/science/microsites/G/great_global_warming_swindle/programme.html [accessed: 20 July 2009].

Cottle, S. 1993. Mediating the environment: modalities of TV news, in *The Mass Media and Environmental Issues,* edited by A. Hansen, Leicester; New York: Leicester University Press; St. Martin's Press.

Deacon, D. 1999. *Researching Communications: A Practical Guide to Methods in Media and Cultural Analysis.* London: Arnold.

Encyclopedia Britannica Online. 2009. *Climate Change-The Global Effects.* [Online]. Available at: www.britannica.com/EBchecked/topic/1517364/Climate-Change-The-Global-Effects [accessed: 28 July 2009].

Fairclough, N. 1995. *Media Discourse.* London; New York: E. Arnold.

Foucault, M. 1972. *Archaeology of knowledge.* Translated by Smith, S. London: Tavistock Publications.

Gunter, B. and Wober, M. 1983. Television viewing and public perceptions of hazards to life. *Journal of Environmental Psychology,* 3 (4), 325–335.

Hall, S. 1992. The west and the rest: discourse and power, in *Formations of Modernity*, edited by H. Stuart and B. Gieben. Oxford: Polity in association with Open University.

Hall, S. 1997. *Representation: Cultural Representations and Signifying Practices.* London: Sage.

Hansen, A. 1993. Greenpeace and press coverage of environmental issues, in *The Mass Media and Environmental Issues,* edited by A. Hansen. Leicester; New York: Leicester University Press; St. Martin's Press.

Hansen, A. 1998. *Mass Communication Research Methods.* New York: New York University Press.

Iedema, R. 2001. Analysing film and television: a social semiotic account of hospital: an unhealthy business, in *Handbook of Visual Analysis,* edited by T. Van Leeuwen and C. Jewitt. London; Thousand Oaks, Calif.: Sage.

Ingram, D. 2000. *Green Screen: Environmentalism and Hollywood Cinema.* Exeter: University of Exeter Press.

Lindahl Elliot, N. 2006. *Mediating Nature.* London; New York: Routledge.

Macdonald, M. 2003. *Exploring Media Discourse.* London: Arnold.

Murray, R.L. and Heumann, J.K. 2009. *Ecology and Popular Film: Cinema on the Edge.* New York: SUNY Press.

Nisbet, M. and Myers, T. 2007. The polls-trends twenty years of public opinion about global warming. *Public Opinion Quarterly*, 71 (3), 444–470.

Noelle-Neumann, E. 1974. The spiral of silence: a theory of public opinion. *Journal of Communication*, 24 (2), 43–51.

Noelle-Neumann, E. 1993. *The Spiral of Silence: Public opinion – Our Social Skin* (2nd ed.). Chicago: University of Chicago Press.

Parlour, J.W. and Schatwow, S. 1978. The mass media and public concern for environmental problems in Canada 1960–1972. *International Journal of Environmental Studies,* 13 (1), 9–17.

Protess, D.L., Cook, F.L., Curtin, T.R., Gordon, M.T., Leff, D.R., McCombs, M.E., and Miller, P. 1987. The impact of investigative reporting on public opinion and policymaking targeting toxic waste. *Public Opinion Quarterly,* 51 (2), 166–185.

Rose, G. 2001. *Visual Methodologies: An Introduction to the Interpretation of Visual Materials.* London; Thousand Oaks, Calif.: Sage.

Shanahan, J. 1993. Television and the cultivation of environmental concern: 1988–1992, in *The Mass Media and Environmental Issues,* edited by A. Hansen. Leicester; New York: Leicester University Press; St. Martin's Press.

Silverstone, R. 2005. The sociology of mediation and communication, in *The SAGE Handbook of Sociology,* edited by C. Calhoun, C. Rojek, and B. Turner. London: Sage.

Smith, J. (ed.) 2000. *The Daily Globe: Environmental Change, the Public and the Media.* London: Earthscan.

The Age of Stupid (dir. Franny Armstrong, 2009).

The Age of Stupid Net. 2009. *Crowd funding FAQ.* [Online]. Available at: www.ageofstupid.net/node/1253/ [accessed: 09 July 2009].

Thompson, J. 1995. *The Media and Modernity: A Social Theory of the Media.* Cambridge: Polity Press.

Wilson, K.M. 2000. Communicating climate change through the media. Predispositions, politics and perceptions of risk, in *Environmental Risks and the Media,* edited by S. Allan, B. Adam and C. Carter. London: Routledge, 201–217.

Wober, M. and Gunter, B. 1985. Patterns of television viewing and of perceptions of hazards to life. *Journal of Environmental Psychology,* 5 (1), 99–108.

Communicating the Political Act of Switching off the Light: Mediating Citizen Action through 'Boundary Acts' in the Earth Hour and Vote Earth Global Media Events

Paul McIlvenny

Introduction

"Earth Hour" began in 2007 in Sydney, Australia, and only two years later, on 28 March 2009, it was proclaimed a successful global event that took place in over 80 countries and 4000 cities.[1] As a material discursive 'eventful protest' (della Porta 2008) or 'eventspace' (Volkmer and Deffner 2010), "Earth Hour" focuses primarily on using new media to communicate and stimulate (global) citizen awareness, participation and solidarity in mitigating climate change. The "Earth Hour" campaign has the goal to persuade citizens to take part in a collective act at an appointed date/time in order to raise environmental awareness. However, in "Vote Earth", the sister campaign to "Earth Hour" in 2009, the discourse of representative democracy was deployed to help constitute an imaginary global electorate (or phantom public). Switching off the lights or not was analogous to voting in a two-party election in which earth and climate change were the only two contenders.[2] According to the campaign, an individual's act of voting by switching off the lights was the means by which he or she expresses support or preference. The idea that one could mediate politics through the mediated act of switching off the lights became the focus of attention in a variety of participatory media, one of which was uploaded videos to YouTube.

Since "Earth Hour" is primarily about responding to climate change and it uses the media to propagate its message, then there are some questions that are immediately relevant. What are the main discourses of climate change that are mediated and circulated? How are the issues and the call to participate in the event mediated? How are people called upon to act? Do they resist this call, and if so,

---

1   See: http://www.earthhour.org. The 2010 event was rather low key, which may have been a result of despondency after the fiasco of the COP15 meeting in Copenhagen, but the "Earth Hour" which took place on 26 March 2011 was more upbeat.

2   The global slogan in English was "Your Light Switch Is Your Vote".

then how? This chapter applies some of the major concepts and methods developed in the field of mediated discourse analysis (Norris and Jones 2005, Scollon 2001a, Scollon 2001b, Scollon and Scollon 2004) in order to analyse how discourses of sustainability and climate change are (re)mediated in these global media events and through which significant mediated actions this takes place. Scollon argues that although mediated discourse analysis takes a strong interest in discourse, it also takes as one of its central tasks to explicate and understand how the broad discourses of our social life are engaged (or not) in the moment-by-moment social actions of social actors in real time activity (2001b: 160). Since all action is mediated (Wertsch 1994, Wertsch 1995), Scollon (2001b: 3) contends that we need to focus "on social actors as they are acting because these are the moments in social life when the Discourses in which we are interested are instantiated in the social world as social action, not simply as material objects."

This chapter examines the discursive articulation of the 'boundary act' of the click/switch as a significant mediated action, which is part of the nexus of practices in which the click/switch is assembled and its meaning negotiated as a political act in relation to climate change. A 'boundary act' is analogous to the concept of 'boundary object' developed by Star & Greismier (1989) to account for *objects* that serve as an interface between different communities of practice, e.g. between different scientific and expert communities. A boundary object is a stable, but flexible entity shared by several different communities, but viewed or used differently by each of them.[3] It facilitates coordination and the management of uncertainty. I suggest that a boundary act is an *action* that serves as an interface between different nexus (or communities) of practice; however, the act is understood and practised differently across the nexus. This chapter also analyses some of the key visualisations of the "Earth Hour" event (for example, in the official videos promoting the event) and the discursive framing and contestation (for example, in YouTube video blogs) of the significance of the anticipated actions, such as switching off the lights (or acting to mitigate global warming).

Climate Change: Discourse, Media and Public

Although in the literature on climate change communication and discourse (see Moser 2010, Nerlich *et al.* 2010, for recent reviews) there is a strong focus on analysing media texts that represent climate change, and some attempt has been made to locate the media audience and their perceptions of the debate, there is little understanding yet of how climate change discourses come to have effects – what they mediate – in specific sites of engagement, and just what ordinary people

---

3   For example, the 350.org campaign uses the 350 ppm CO_2 limit as a stable, iconic figure for many diverse communities to organise around. See Shackley and Wynne (1996) for an early discussion of the relevance of this concept and the more transient 'boundary-ordering device' in understanding uncertainty in climate change science and policy.

'do' when called upon to act by a campaign. One of the key issues with respect to environmental communication today is how discourses of climate change and social action are taken up, circulated and responded to. Hence, we can ask what notion of 'public' or 'audience' is invoked by climate discourses?[4] For example, Lewis (2008) samples green lifestyle television in the UK to determine whether or not they inculcate new forms of civic responsibility and citizenship. Lewis looks into the "ways in which contemporary lifestyle TV and lifestyle culture might be articulated to and generative of modes of popular civic politics that speak to more progressive models of citizenship and activism, suggesting a more complex and contested relationship between popular media culture, public/governmental concerns and ordinary, everyday practices of consumption" (227). One of the key issues with my approach to "Earth Hour" is how particular discourses of climate change are circulated at particular sites of engagement (e.g. in videos and on YouTube), and thus what sorts of mundane and spectacular effects (e.g. "tipping points", see Russill 2008) on everyday life and social action they might have.

Data

In order to document the "Earth Hour"/"Vote Earth" event in 2009, posters, leaflets, newspaper articles, blogs, websites, online Flikr photo and YouTube video archives and online news archives were collected in the weeks prior to and after the event on 28th March. Out of these diverse and sometimes competing representations of the event, I am particularly interested in the diversity of responses that everyday people made, especially in the form of video blogs (YouTube).[5] In this chapter, rather than do interview surveys with potential audiences of Earth Hour, I suggest that we can use the variety of YouTube responses incited by the "Earth Hour" campaign and website to gain access to some of the sites of engagement, discursive genres and modes of acting globally that framed participation in "Vote Earth". Moreover, we may be able to find some answers to the puzzle of how discourses of climate change and social action are

---

4   There are many studies of audience versus public connection (Couldry *et al.* 2007, Couldry and McCarthy 2004, Livingstone 2005). It is clear that participation in such a protest event creates a public (Barry 2001, della Porta 2008, Marres 2008, Marres 2009). We might conjecture that Earth Hour assembles a material public (Asdal 2008, Latour 2004, Latour 2005, Latour and Sánchez-Criado 2007, Latour and Weibel 2005) or performs a new environmental citizenship (Dobson and Bell 2005, Jagers 2009, Smith and Pangsapa 2008, Whitmarsh et al. 2011a, Whitmarsh et al. 2011b).

5   There have been academic studies of different aspects of YouTube, which went online in 2005, and video archives (Buckingham et al. 2007, Buckingham and Willett 2009, Buckingham et al. 2011, Burgess and Green 2009, Gehl 2009, Harley and Fitzpatrick 2009, Hilderbrand 2007, Jarrett 2008, Lange 2007a, Lange 2007b, Lange 2007c, Snickars and Vonderau 2009).

taken up, circulated and responded to. For example, the many short YouTube videos that were uploaded give us an audiovisual *representation* of moments of switching off, which are indicative of the diversity of experiences of "Earth Hour", and of the circulation of specific public and private discourses of climate change. Even though they are mediated and edited fragments, these are natural, *endogenous* forms of participation that are assembled both as part of a topology of climate discourse activism and as a nexus of practice.

The campaign videos: constructing the boundary act in "Earth Hour"
and "Vote Earth"

The WWF was responsible for at least five short campaign videos, which were uploaded to YouTube. These official videos used animation, documentary footage, and celebrity endorsements to different degrees. All of them had a persuasive function. I will refer to them as follows:

- Earth Hour (EA)
1. Documentary[6]
- Vote Earth (VE)
1. Persuasive animation[7]
2. Australian celebrity endorsements[8]
3. Kids' version[9]
4. Teenagers' version

Rather than asking people to donate money as an act of charity, the "Earth Hour" event calls upon everyone to make a statement concerning global warming by switching off the lights at the designated hour.[10] Thus, it demands an unequivocal binary act, to act-at-a-distance – using both the domestic electricity supply and the regional grid – through a discursive material technology of the binary on/ off (or vote/not vote).[11] In the campaign videos, the boundary act of switching is

---

6   http://www.youtube.com/watch?v=1CRs-7lRlPo. This video has had over one million views. It was uploaded on the 11th November 2008. In July 2009, a second post-event video was uploaded: "Earth Hour's Vote Earth. It's Time To Show Where You Stand": http://www.youtube.com/watch?v=gNxvNm7rqBQ.

7   http://www.youtube.com/watch?v=f2gfq2-ge5U. This video has had over one hundred thousand views and was uploaded on 23rd February 2009.

8   http://www.youtube.com/watch?v=BHeb-GYjny0

9   http://www.youtube.com/watch?v=b-Z-DNe-8sc

10   The campaign videos requested a simple material symbolic act, but not a real sacrifice nor a serious demand to save energy on the part of citizens, institutions and corporations.

11   Voting 'no' is *not* an explicit possibility in the campaign. It is the absence of a vote for earth that has consequences.

visualised or implied in different ways. For example, the screen goes black, or a metonymic light switch appears and is switched off. Alternatively, on a dynamic map of the globe on which aggregate light intensity is plotted, a city (Sydney) goes dark. Or clips are shown of footage of actual buildings or tourist sites, where the lights were visibly switched off (in past events in 2007 or 2008). Clearly, this campaign assumes the pervasiveness of the embodied technology of switching and the ease of access and control that ordinary citizens have to this mundane, domesticated technology. Of course, not everyone in the world has convenient access to electricity, let alone lights and appliances to switch off. But then, they are not the ones who necessarily need to save energy to help the living conditions of the world's poor.

On the other hand, the consequences of not switching off are visualised in terms of a Manichean struggle between good 'earth' and evil 'global warming'. Repeatedly, the video campaign construes climate change as a character in a moral drama between the good 'planet'/us and the bad 'global warming'. In several videos, the space of the image or the space around the talking head is coded in these terms. In the classic spatial configuration of given/new in an image (Kress and Leeuwen 2006), left is good, right is bad (sometimes, without apparent contradiction, it is reversed). The good image of earth is a cool green and blue globe; the bad image of earth/climate change has a warm orange or red hue. This symbolism appeared in the two official videos discussed below, VE1 and VE2.

In the animation video (VE1), the narrator states "the world's first global election is taking place an election between earth and global warming". As the narrator says "earth", a flat projection of the earth turns into a 2-D representation of the globe, at which point a reddish earth emerges from behind the green earth. As the narrator says "and global warming", this earth moves out from its occluded position to appear in a left/right contrast pair with "VS" between the two possible states of earth. The two globes are turning on their axes, but predominantly the 'good' earth shows Europe/Africa and the 'bad' earth shows North America/Pacific Asia, which might represent the regions most responsible for global warming.

In a second video (VE2), a talking head celebrity asks rhetorically: "if there was an election between earth and global warming who would you vote for." Once again, as he says "earth" a green planet appears by his left shoulder, and as he says "global warming" an orange planet appears by his right shoulder. The drips suggest the earth is melting under the conditions of global warming. The campaign, therefore, promotes a shift towards seeing the planet's climate as a specific, autonomous entity (a cosmopolitan actant) requiring our support and allegiance, and the polarisation of two actors within a discourse of conflict or 'climate war' (Dyer 2008).

The deliberative choice *not* to switch off the lights at the designated time becomes a morally accountable act, a virtual vote for climate change and its morally undesirable consequences, both local and global. Thus, the discourse of the campaign insists that their call is global, that everyone is addressed by the call, that the obligation to act is shared and mutual, that the vote is a legitimate (though symbolic) one, and that not voting has consequences. Additionally, the organisers promise to count or aggregate the distributed acts of clicking/voting and the results of the 'vote' will be (and indeed were) presented at the COP15 meeting in Copenhagen in December 2009 as a supplement to the call from global civil society for governments to urgently make 'policy that matters' at the international level. Thus, there is an assurance that voting will have an effect.

But not everyone was impressed with the campaign, and some decided to respond and to register their dissent and/or hinder the circulation of this mediated discourse. For example, there was a sceptical reception of the event in terms of its supposed effect on mitigating anthropogenic climate change. And there was a rather more strident reaction to its insistent call to act. Although the creators of the website campaign for 'Vote Earth' discursively polarised the debate around the simple analogy of <turn off the light> = <vote for good earth>; <not turning off the light> = <vote for bad global warming>, some websites and video blogs on YouTube countered this discursively with an alternative analogy:

- <turn on all the lights> = <vote for civilisation>
- <turn off the light> = <vote for barbarism or stupidity>.[12]

I return to the dissenting video blogs in more detail below.

It is clear that there was a massive communications infrastructure of discourse and frames to synchronise the collective performance of the event and to constitute a global, 'concerned public'. New media technologies were appropriated and harnessed by WWF/"Earth Hour" to constitute a global public for the event. In the campaign, there was an intense drive to visualise and spectacularise the effects of the simple act of switching off the lights. Many websites, online video archives and blogs, both official and personal, contain photos and video images of the activities of participants in "Earth Hour" and the before/after effect of switching lights off at the designated local time. Moreover, in the lead up to the Earth Hour event and after, there was a heavy emphasis on doing discursive 'memory work' to witness or memorialise the hour. The main website for "Earth Hour" called on those who participated to document and archive their experiences on online video archives, such as YouTube.

YouTube: mediating participation in the campaign

In the last few years, the video media distribution website YouTube has attained a special role in the democratisation of the global distribution and archiving of moving images (Burgess and Green 2009, Gehl 2009). In fact, there is a dedicated "Earth Hour" group in the YouTube universe to which uploads could be tagged

---

12   Wikipedia reports that one libertarian think tank, the Competitive Enterprise Institute, proposed an alternative celebration of "Human Achievement Hour" to celebrate the advancement of human prosperity. http://en.wikipedia.org/wiki/Earth_hour.

"Earth Hour Global".[13] For the rest of this paper, I focus on the more personal YouTube videos – video blogs or vlogs – by newbies and regular YouTubers. A typical vlog on YouTube consists of a head shot of a blogger facing their web camera, which is placed behind or on top of their computer monitor. From the *mise-en-scène* of the video, we can determine that the blogger is often at home in the privacy of their bedroom or study. The blogger talks to the camera in a confessional mode. Often the video that is uploaded is dialogic, in that it engages with other YouTube videos or YouTubers' comments. Thus, eye gaze, gesture, and verbal style may all play important roles in the self-presentational rhetoric of the blogger (Harley and Fitzpatrick 2009).

When they are uploaded, videos can be categorised and self-tagged using an emic set of social classifications (otherwise known as a 'folksonomy'), which play an important role in the ideological space created by the search engine that users of YouTube use to find videos (Halavais 2009), and, hence, in who might view the videos and the interpretation that they may make of them. From a small corpus of eleven personal videos uploaded to YouTube about "Earth Hour"/"Vote Earth" in 2009, which were selected after a random search because of their representativeness of different types of responses, the following tags were given (including any spelling mistakes)[14]:

1. "Vote Earth Hour WWF voteearth earthhour earthour 2009 animation climate change non-profit lightsoff lights off fix you global warming environment carbon election light switch billion world Licht aus Welt Erde umwelt XXXXX sono XXXXXX copenhagen kyoto"[15]
2. "grassroots outreach political commercial coco show youtube subscriber subscribe XXXXXXX ontario XXX XX kent canada earth"
3. "voteearth09 XXXXXX XXXXXXXX earth hour me singing say it again marie digby singapore singapura dark house cat talking"
4. "dark lights earth hour environment 10 things no XXXXXX"
5. "australia"

---

13   http://www.youtube.com/groups_videos?name=earthhourglobal

By the end of April 2009, there were 622 videos, 940 members subscribed and 22 discussions in this YouTube group channel.

14   In comparison, the tags for the "Earth Hour" video EA1 were: "earth hour climate change earthhour earthhour2009 vote voteearth global warming globalwarming wwf climatechange Furtado Cate Blanchett Branson Shepard Fairey lights out lightsout world 2009 environment non-profit carbon". For the "Vote Earth" video VE1, they were: "Vote Earth Hour WWF voteearth earthhour earthour 2009 Cate Blanchett Coldplay Shepard Fairey animation climate change non-profit lightsoff lights off fix you global warming environment carbon election light switch billion world".

15   XXXXXXX marks a tag that has been removed because it could identify the vlogger. Although the YouTube site is public and the videos uploaded are freely available, I have still anonymised the identity of the vloggers in the data presented here.

6. "earth hour climate change environment energy vlog XXXXXXXXXXXX will turn off his lights tonight"
7. "The Lights On Hour 2009 earth hour global warming climate change scam erath climnate off dark"
8. "地球温暖化　アースアワー　地球の日　Global Warming Earth Hour Day"
9. "am against earth hour XXXXXXXXXX vote technology humanity"
10. "am against earth hour response part ii XXXXXXXXXX technology electrical dip"
11. "earth hour 2009 zealand con sham swindle hypocrisy global warming climate change energy ration"

Certain patterns can be seen across this small corpus: the Earth Hour event and the topic of climate change feature prominently, of course. An indication of the boundary act of switching off the lights features in some of the video tag lists above, but not in others (for example, not in those that were critical of the event). The location of the vlogger is sometimes indicated in the corpus, e.g. Singapore, Canada, Australia, New Zealand, which illustrates the global response found on YouTube. Critics label the event as a "scam" or "sham", or indicate their stance clearly, e.g. "am against…" Several vloggers above use the syntagmatic dimension afforded by the list to write grammatical clauses or phrases indicating action, e.g. "me singing say it again…", "am against earth hour", "vote technology", and "XXXXXXXX will turn off his lights tonight". These initial categorisations led me to research further the actual video responses to Earth Hour, particularly Vote Earth, by YouTubers.

Although the YouTube videos about Earth Hour are often generic in format and style and authorship is restricted to a particular segment of the population, YouTube provides us with an archive of ordinarily occasioned and organised responses to a global event, which would otherwise be hard to obtain, e.g. by a survey or an ethnography in domestic homes. For instance, the many short YouTube videos that were uploaded do give us *audiovisual and highly mediated representations* of moments of switching off. They demonstrate the diversity of experiences of "Earth Hour" and of the circulation of specific public and private discourses of climate change.

The online YouTube video archive concerning "Earth Hour" includes uploaded videos which are categorisable into at least three main genres: a) pre-event information slot; b) 'live' recording during the event; and c) critical responses.[16] Out of the 600 or more uploaded videos, the predominant genre was the pre-event info slot, closely followed by the 'live' recording. There were less than twenty

---

16   A fourth genre not discussed here would be the official compilations or documentations of a city or corporate effort, e.g. Sydney or Beijing city landmarks.

critical responses. Given space restrictions, only one video example is given in more detail below (with a transcript).[17]

Pre-event info video blogs

In the pre-event information slot, a video is usually anticipatory: it prefigures particular actions and stances before the event takes place. For example, actions by the self may be highlighted. The actual act of switching off the lights is often mitigated: it is just a simple act. A vlogger may also practice switching off the lights for fun. Recommendations about what to do may also be given. These videos are generally advocating the event, passing on relevant information and indicating that the vlogger is also going to participate, and thus it is performed as a shared virtual event (even though the vlogger is usually alone in their video). Many pre-event videos affirmatively recycle the discourse of "Earth Hour" and anticipate particular actions they will take which exemplify appropriate conduct in the event and the disruption it causes to their everyday practices.

The 'Live' Video Blogs

In the second category of video blogs, the videos feature the blogger 'live' during the "Earth Hour" event. The vlogger is often in a darkened room – already 'doing' "Earth Hour". Sometimes the lights are switched off or on to mark the beginning or end of the local "Earth Hour". Vloggers often identify their local time and location, which places them in a local milieu, but also within the demonstrable symbolic time frame, e.g. "It's um eight forty one pee em in Singapore and erm we've had out lights off for about ten minutes now erm it's earth hour erm (   ) and erm celebrating it yea::h". The embodied vlogger is both living 'the moment' and orienting to the construction of an archival document for future viewing. The liveness of the event is staged using a variety of verbal means and visual and material props. With a sense of immediacy and naturalistic presence, bloggers may show – by panning their web camera or by giving a guided tour of their location, for example – proof that their house/flat or locality is also darkened. The fact that certain electronic devices are still on in the presence of the vlogger is accounted for, e.g. laptop on batteries, to reassure potential viewers. Other activities may be engaged in, such as playing with the cat, singing a song, listing 'green' things to do, etc.. Thus, the vlogger can demonstrate an appropriate attitude towards the boundary act and an awareness of sustainable discourses and practices.

---

17   This choice is to highlight some of the limitations and inherent contradictions of the campaign as understood by YouTubers who oppose Earth Hour on principled grounds.

The Critical/Sceptical Video Blogs

Although many vloggers supported and promoted the campaign, some responded negatively to the call to participate in "Earth Hour" or "Vote Earth". Some were weakly supportive but sceptical; others were argumentative or insulting. A few illustrated their point by flaunting material artefacts that for them are symbolic of consumption, or by showing that they had *not* done the boundary act required, i.e. to switch off their lights.

In the Extract (see Figure 4.3.1), we have an example of a vlogger who is resistant to the "Earth Hour" message.[18] In this case, as in others like it, he positions himself in receipt of what he interprets as an imperative, namely the request to switch off the lights, interpolated by an instruction or requirement from an outside yet unnamed authority (lines 1–4). Clearly, this blogger is suspicious of the voluntary nature of the gesture, and he suggests in the description attached to the video that it prefigures a drastic governmental intervention.

In this indignant video blog, the vlogger intercuts a slideshow of still photographs (lines 7–12) indexing poverty, destitution and distant suffering, which draw upon a politics of pity (Boltanski 1999). He performatively appropriates the classic topos of denunciation – his apparent anger at the inequality in the world between the Global South and the minority in the rich North – to argue that we ought to celebrate how our lives have been transformed by advanced technology – "electrical appliances" – that can be turned on. In terms of the reporting of the suffering of others in the mass media news, Chouliaraki (2006) argues that we generally find three types: adventure, emergency and ecstatic news. Here we find a cursory mode of adventure news. To conclude his short blog (lines 13–19), he documents visually with a web camera pan around the room that he has, in fact, switched *on* all his lights. He marks this brazenly profligate use of energy as a sign of civilisation or modernity (lines 14–16), and thus obviates the moral imperative that he ought to be more sustainable in his use of carbon-generating resources. Thus, he is directly challenging a key assertion in the discourse of "Vote Earth", in which switching off is a vote for earth. Instead, he has deliberately switched on to valorise an "energy intensive lifestyle" (lines 14–15), which makes us different from (and better than) those who lack such a lifestyle.

Conclusion

From an analysis of the official videos, as well as the participatory and responsive YouTube video blogs, we can see how the discourse of "Earth Hour" and "Vote Earth" (and the boundary act of switching off) circulates and is relayed or resisted,

---

18   Transcript conventions are based on conversation analysis's standard conventions (Have 2007) with additional markings for image placement, e.g. "@" marks when an image occurs within the stream of talk.

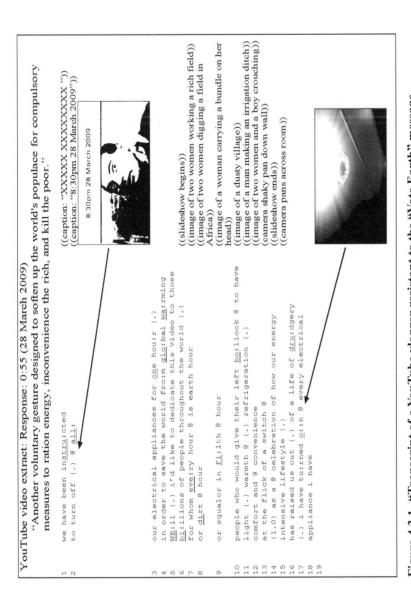

YouTube video extract: Response: 0:55 (28 March 2009)

"Another voluntary gesture designed to soften up the world's populace for compulsory measures to ration energy, inconvenience the rich, and kill the poor."

((caption: "XXXXX XXXXXXXX "))
((caption: "8:30pm 28 March 2009"))

8:30pm 28 March 2009

1   we have been instru:cted
2   to turn off (.) @ all:

3   our electrical appliances for one hou:r (.)
4   in order to save the world fro:m glo:bal wa:rming
5   WE:ll (.) i'd like to dedicate this video to those
6   bi:llions of people throughout the world (.)
7   for whom eve:ry hour @ is earth hour
8   or dirt @ hour

9   or squalor in fi:lth @ hour

10 people who would give their left bo:llock @ to have
11 light (.) warmth @ (.) refrigeration (.)
12 comfort and @ convenience
13 at the flick of a switch @
14 (1.0) as a @ celebration of how our energy
15 intensive lifestyle (.)
16 has raised us out (.) of a life of dru:dgery
17 (.) i have tu:rned o::n @ every electrical
18 appliance i have
19

((slideshow begins))
((image of two women working a rich field))
((image of two women digging a field in Africa))
((image of a woman carrying a bundle on her head))
((image of a dusty village))
((image of a man making an irrigation ditch))
((image of two women and a boy crouching))
((camera shaky pan down wall))
((slideshow ends))
((camera pans across room))

Figure 4.3.1 "Transcript of a YouTube vlogger resistant to the "Vote Earth" message

and, therefore, that what it means to participate in such an event is negotiable. The official videos attempt to inculcate a more durable 'global citizen', one who is concerned with the environment/climate and whose anticipatory mediated actions in relation to it are prefigured by the event (see Carvalho 2007). The discursive innovation of this event is to link practices of democratic participation with everyday individual acts – the mediated act of switching off the lights – to fashion a discursive material public who act collectively in the face of global warming and climate change. To some extent, those who participate and document this online (for instance, in YouTube) reproduce the upbeat sustainability discourses in relation to mitigating climate change. Those who disagree, tend to focus critically on the incitement to participate as an act of voting for earth. Some disagree flatly with the claim that there is a phenomenon called climate change; others just point out the irony of the spectacle of the event or disagree with the argument that switching off lights is in any way helpful.

One of the key goals of the official "Earth Hour" campaign was to fix temporarily the meaning of a simple boundary act, namely switching off the light, and to establish it as a flexible, viral cultural tool for social activism on climate change. But as we know from studies of communication, it is not easy to fix the meaning of a social action, esp. when it is interactional and therefore negotiable. We can ask, therefore, what potential is there in looking at specific types of boundary acts as glocal mediational means, such as the switching gesture? The answer might lie in the emergence of new global 'techniques of the body' (and of talk) as part of a move towards green governmentality or environmentality. Bayert (2007: 209–210) points towards the emergence of global (political) techniques of the body as part of an emerging global governmentality. Bayert understands the 'conducts of life' engendered by globalisation as an encounter between techniques of domination over others and techniques of the self, in which consumption and the practices of material culture are crucial.[19] He shows how global social institutions regulate the movements of the body, for example in factories, the global Scout movement, flashmobs, and international hotels and airports. In the case of "Earth Hour", the gesture of switching off the electric light is globalised as a shared political technique of the body in an event of appropriation.[20]

With respect to effecting social change, what is also critical, but cannot be answered without an ethnographic study of post-event conduct, is whether or not there has been a shift in discourses (or the order of discourse) *and* practices as a result of this event. Is the coupling of a local act to discourses of sustainability, electoral democracy and moral citizenship a new configuration or nexus of practice? Will it be sustainable and flourish? Will it be effective? At this point,

---

19   See Paterson and Stripple (2010) for an investigation into how the "conduct of carbon conduct" is managed in a variety of approaches to reducing individual carbon emissions.

20   We could ask, also, what other boundary acts might constitute a more effective glocal technique of the body for "Earth Hour" and similar global media events?

we can only speculate whether or not Earth Hour's focus on a simple material symbolic gesture, with the implication that there is a single interpretation of that gesture across all cultures (not true for boundary acts), may ultimately be unsuccessful or irrelevant in the long time struggle over climate change. This may be because it will inadvertently sustain unsustainable concepts of individual action (and sacrifice).[21]

References

Asdal, K. 2008. On politics and the little tools of democracy: a down-to-earth approach. *Distinktion: Scandianvian Journal of Social Theory*, 16, 11–26.

Barry, A. 2001. Demonstrations: sights and sites, in *Political Machines: Governing a Technological Society*, London: The Athlone Press, 175–198.

Bayart, J.-F. 2007. *Global Subjects: A Political Critique of Globalization*. Cambridge: Polity Press.

Boltanski, L. 1999. *Distant Suffering: Morality, Media and Politics*. Cambridge: Cambridge University Press.

Buckingham, D., Pini, M. and Willett, R. 2007. 'Take back the tube!': the discursive construction of amateur film and video making. *Journal of Media Practice*, 8(2), 183–201.

Buckingham, D. and Willett, R. (Eds.) 2009. *Video Cultures: Media Technology and Everyday Creativity*. Basingstoke: Palgrave Macmillan.

Buckingham, D., Willett, R. and Pini, M. 2011. *Home Truths? Video Production and Domestic Life*. Ann Arbor: The University of Michigan Press.

Burgess, J. and Green, J. 2009. *YouTube: Online Video and Participatory Culture*. Cambridge: Polity Press.

Carvalho, A. 2007. Communicating global responsibility? discourses on climate change and citizenship. *International Journal of Media and Cultural Politics*, 3(2), 180–183.

Chouliaraki, L. 2006. *The Spectatorship of Suffering*. London: Sage.

Couldry, N., Livingstone, S. and Markham, T. 2007. *Media Consumption and Public Engagement: Beyond the Presumption of Attention*. Basingstoke: Palgrave Macmillan.

Couldry, N. and McCarthy, A. (Eds.) 2004. *Mediaspace: Place, Scale and Culture in a Media Age*. London: Routledge.

della Porta, D. 2008. Eventful Protest, Global Conflicts. *Distinktion: Scandinavian Journal of Social Theory*, 17, 27–56.

---

21 In addition to the lull in activity after the disastrous COP15 meeting in December 2009, this may be one of the reasons why the "Vote Earth" campaign was dropped in 2010. I enquired, but no response from the communications director of the campaign was forthcoming.

Dobson, A. and Bell, D. (Eds.) 2005. *Environmental Citizenship: Getting from Here to There*. Boston, MA: MIT Press.

Dyer, G. 2008. *Climate Wars*. Canada: Random House.

Gehl, R. 2009. YouTube as archive: who will curate this digital *wunderkammer*? *International Journal of Cultural Studies*, 12(1), 43–60.

Halavais, A. 2009. *Search Engine Society*. Cambridge: Polity Press.

Harley, D. and Fitzpatrick, G. 2009. Creating a conversational context through video blogging: the case of geriatric1927. *Computers in Human Behavior*, 25(3), 679–689.

Have, P. t. 2007. *Doing Conversation Analysis: A Practical Guide (2nd edition)*. London: Sage.

Hilderbrand, L. 2007. YouTube: where cultural memory and copyright converge. *Film Quarterly*, 61(1), 48–57.

Jagers, S. C. 2009. In search of the ecological citizen. *Environmental Politics*, 18(1), 18–36.

Jarrett, K. 2008. 'Beyond broadcast yourself™': the future of YouTube. *Media International Australia*, 126, 132–244.

Kress, G. and Leeuwen, T. 2006. *Reading Images: The Grammar of Visual Design (2nd edition)*. London: Routledge.

Lange, P. G. 2007a. *Commenting on Comments: Investigating Responses to Antagonism on YouTube*. Society for Applied Anthropology Conference, Tampa, Florida, 31 March 2007. Available at: <http://anthropology.usf.edu/cma/Lange-SfAA-Paper-2007.pdf> [accessed: 14 April 2011].

Lange, P. G. 2007b. Publicly private and privately public: social networking on YouTube. *Journal of Computer-Mediated Communication* [Online], 13(1). Available at: <http://jcmc.indiana.edu/vol13/issue1/lange.html> [accessed: 14 April 2011].

Lange, P. G. 2007c. The vulnerable video blogger: promoting social change through intimacy. *The Scholar and Feminist Online* [Online], 5(2). Available at: <http://www.barnard.edu/sfonline/blogs/printpla.htm> [accessed: 14 April 2011].

Latour, B. 2004. Whose cosmos, which cosmopolitics? comments on the peace terms of Ulrich Beck. *Common Knowledge*, 10(3), 450–462.

Latour, B. 2005. *Reassembling the Social: An Introduction to Actor-Network-Theory*. Oxford: Oxford University Press.

Latour, B. and Sánchez-Criado, T. 2007. Making the '*res* public'. *Ephemera: Theory & Poltics in Organization* [Online], 7(2), 364–371. Available at: <http://www.ephemeraweb.org> [accessed: 14 April 2011].

Latour, B. and Weibel, P. (Eds.) 2005. *Making Things Public: Atmospheres of Democracy*. Cambridge, Mass.: The MIT Press.

Lewis, T. 2008. Transforming citizens? green politics and ethical consumption on lifestyle television. *Continuum*, 22(2), 227–240.

Livingstone, S. (Ed.) 2005. *Audiences and Publics: When Cultural Engagement Matters for the Public Sphere*. Bristol: Intellect.

Marres, N. 2008. The making of climate publics: eco-homes as material devices of publicity. *Distinktion: Scandinavian Journal of Social Theory*, 16, 27–46.

Marres, N. 2009. Testing powers of engagement: green living experiments, the ontological turn and the undoability of involvement. *European Journal of Social Theory*, 12(1), 117–133.

Moser, S. C. 2010. Communicating climate change: history, challenges, process and future directions. *Wiley Interdisciplinary Reviews: Climate Change*, 1(1), 31–53.

Nerlich, B., Koteyko, N. and Brown, B. 2010. Theory and language of climate change communication. *Wiley Interdisciplinary Reviews: Climate Change*, 1(1), 97–110.

Norris, S. and Jones, R. (Eds.) 2005. *Discourse in Action: Introduction to Mediated Discourse Analysis*. London: Routledge.

Paterson, M. and Stripple, J. 2010. My space: governing individuals' carbon emissions. *Environment and Planning D: Society and Space*, 28(2), 341–362.

Russill, C. 2008. Tipping point forewarnings in climate change communication: some implications of an emerging trend. *Environmental Communication: A Journal of Nature and Culture*, 2(2), 133–153.

Scollon, R. 2001a. Action and text: towards an integrated understanding of the place of text in social (inter)action, mediated discourse analysis and the problems of social action, in *Methods of Critical Discourse Analysis*, edited by R. Wodak and M. Meyer. London: Sage, 139–183.

Scollon, R. 2001b. *Mediated Discourse: The Nexus of Practice*. London: Routledge.

Scollon, R. and Scollon, S. W. 2004. *Nexus Analysis: Discourse and the Emerging Internet*. London: Routledge.

Shackley, S. and Wynne, B. 1996. Representing uncertainty in global climate change science and policy: boundary-ordering devices and authority. *Science, Technology & Human Values*, 21(3), 275–302.

Smith, M. J. and Pangsapa, P. 2008. *Environment and Citizenship: Integrating Justice, Responsibility and Civic Engagement*. London: Zed Books.

Snickars, P. and Vonderau, P. (Eds.) 2009. *The YouTube Reader*. Stockholm: National Library of Sweden.

Star, S. L. and Griesemer, J. 1989. Institutional ecology, 'translations' and boundary objects. *Social Studies of Science*, 19, 387–420.

Volkmer, I. and Deffner, F. 2010. Eventspheres as discursive forms: (re-) negotiating the 'mediated center' in new network cultures, in *Media Events in a Global Age*, edited by N. Couldry, A. Hepp and F. Krotz. London: Routledge, 217–230.

Wertsch, J. V. 1994. The primacy of mediated action in sociocultural studies. *Mind, Culture and Activity*, 1(4), 202–208.

Wertsch, J. V. 1995. Mediated action and the study of communication: the lessons of L.S. Vygotsky and M.M. Bakhtin. *The Communication Review*, 1(2), 133–154.

Whitmarsh, L., O'Neill, S. and Lorenzoni, I. (Eds.) 2011a. *Engaging the Public with Climate Change: Behaviour Change and Communication*. London: Earthscan.

Whitmarsh, L., Seyfang, G. and O'Neill, S. 2011b. Public engagement with carbon and climate change: to what extent is the public 'carbon capable'? *Global Environmental Change*, 21(1), 56–65.

Climate Activism and the Mass Media: Potentially a Politically Challenging Debate

Anders Danielsen

Introduction

A group of around 1,500 Danish climate activists attempted to close down the Danish coal power plant, Amager Power Station, in a direct action called 'Shut It Down' on 26 September 2009. The activists were trying to address what they considered an urgent need, which was to end the use of coal in Denmark immediately, thus reducing the nation's contribution to global warming.[1] The activists characterised this form of direct action as confrontational, non-violent civil disobedience. However, a strong police presence prevented the activists from shutting down the plant, which is owned by the state-owned Swedish power company, Vattenfall. The confrontation between the police and the activists, where activists tried to force their way past the police line, resulted in the police using their batons to prevent them from entering the grounds of the power station. The confrontation and the activists' protest march to the power station were followed closely, first hand, by a large group of journalists and press photographers. The direct action came to an end when the police arrested a large number of activists and the other activists were forced to abandon the area around the power plant.[2]

In the months leading up to the event, the activists had openly announced the action (Borking 2009a, dr.dk 2009a, Modkraft.dk 2009c), a strategy also used by climate activists during COP15, the United Nation's Climate Change Conference in Copenhagen in December 2009 (Krause 2009). This strategy of openness enabled the activists to have a voice in the Danish mass media in the months leading up the actual action, as opposed to only on the day of the action. News coverage in the mass media, however, was still much more extensive on the day of the action and in the days leading up to the event.[3] Direct action as a form of protest is a desirable topic for news journalism as it lives up to primary news

---

1   The burning of fossil fuels such as coal, oil, and gas comprise 80% of Denmark's total CO2 emissions (Danish Commission on Climate Change Policy 2010).

2   Based on insights from my observational study.

3   (Borking 2009c, Frøslev 2009, Fyns Stiftstidende 2009, Nørmark 2009, Jp.dk 2009, Politiken.dk 2009, Ritzau 2009b, 2009c, 2009d, 2009e, Schultz 2009) .

values,[4] thus providing climate activists more space in the mass media than they would otherwise normally achieve (de Jong et al. 2005: 6–7, Gabor and Willson 2005: 100–109). The communicative potential of this form of action gave the activists the space to voice their critique of the use of coal by igniting a debate in the Danish mass media, a focal point of this chapter.

This chapter presents a case study of the public debate on the Shut It Down action as depicted in the Danish mass media. The central idea has been to examine how the critical stance of the activists concerning climate change gained a voice in the debate in the Danish mass media and how this debate could potentially have become politically challenging from a hermeneutic interpretative perspective (Højbjerg 2005). A topic or specific act in itself is not inherently political while under debate and must be politicised before becoming a political challenge. Jodi Dean writes:

> Specific or singular acts of resistance, statements of opinion, or instances of transgression are not political in and of themselves. Rather, they have to be politicized, that is, articulated together with other struggles, resistances, and ideals in the course or context of opposition to a shared enemy or opponent (2008: 106).

The formulation of critical statements in a debate concerning a specific topic, i.e. the need to stop the use of coal, thus needs to be politicised by being linked to broader social and political issues through reference to institutional, structural, or abstract normative demands (Dean 2008: 106–15). As a result, the articulation of critical perspectives on a particular topic needs to be politicised for a debate to become politically challenging.

Empirically, this case study[5] is primarily based on a documentary analysis of what the Danish mass media broadcasted, presented, and wrote about the action, focusing, specifically on articles drawn from the major Danish newspapers, the Danish news agency Ritzau, two small left-wing papers, and the TV and radio news broadcasts about Shut It Down by the national Danish public service broadcaster DR.[6] News coverage began with several news stories a couple of months before the action and ended approximately one month after the day of the action. In

---

4   Topicality, essentiality, proximity, sensation and conflict are primary news values in Denmark (Schultz 2006: 57–64).

5   The case study uses a methodological qualitative strategy that has made getting close to the details and grasping the diversity of the empirical data possible. The case study is also an acknowledgement of how knowledge production is context dependent (Flyvbjerg 1991:137-53). The focus of the case study has not been on generalisation, but hopefully the chapter can provide an analytical framework for studying debates in the mass media (Almlund 2007: 265–69).

6   The articles and broadcasts consist of reports, comments, and interviews.

addition to the main case study, I carried out a small observational study[7] on the planning and preparation of the direct action by the activists. The observational study, which also contributed to the empirical foundation of the study, consisted of participating in several internal activist meetings in the months prior to the events surrounding the action and participating on the day of the direct action.

The analysis of the case study is divided into three main sections prefaced by a presentation of the framework used for the analysis. The first section outlines four perspectives articulated during the debate, indicating how news coverage played a vital role in shaping the quantity of space the four perspectives received during the debate. The second section focuses on Vattenfall's response to the activists' critical points and shows how the response followed what Slavoj Zizek has described as post-politics. The final section takes a closer look at the times in the debate when the activists articulated a structurally-grounded critique and how this articulation could have fostered a more politically challenging debate if it had been given a stronger voice in the news coverage.

Four Articulated Perspectives

Based on a dissensus approach to public debate and communication (Haahr 2000), this qualitative analytical study focuses on how the critical messages of the activists are voiced and contested by the different actors and the activists' central opponent in the debate, Vattenfall.

From a hermeneutical perspective, the focus of the examination of the articles and broadcast news has consequently been that of identifying the general perspectives articulated in the debate. This is true both for the perspectives that the activists formulated directly and that, to different degrees, were contested as well as for the perspectives that the direct action as a form of protest sparked. In addition, determining which perspectives became dominant in the debate, and which ones did not, was another analytical focal point. This approach to the debate enabled the identification of four articulated perspectives using a thorough qualitative, analytical systematisation of all the articles and broadcasted news produced on the action. The observational study enabled an in-depth understanding of the activists' communicative strategies and a first-hand account of the day of the direct action, subsequently providing a stronger, more refined approach to identifying these perspectives.

Due to space limitations, the following list provides a summary of each perspective instead of presenting an exhaustive account of all of the details:

---

7 The observation study can best be described as a present observation instead of participatory observation as participating on an equal footing with the activists while also maintaining a reflexive distance is not possible (Almlund 2007: 244–50, Kristiansen and Krogstrup 1999).

- *The role of the market and consumers in the climate issue.* This perspective points to a critique formulated by the activists of the role of the market and people acting as consumers as a solution to the issue of climate change.
- *The climate and energy policies of the Danish government.* This perspective is grounded in the activists' critique of the government's climate and energy policies, where the key focus was upon the use of coal and how eliminating its use for energy production in Denmark is possible.
- *Vattenfall's use of coal.* This perspective points to the discussion between the activists and Vattenfall on the issue of immediately phasing out the use of coal at the Amager Power Station.
- *Civil disobedience as a form of action.* This perspective captures the discussion on the legitimacy of civil disobedience; the tactic used by the activists, as a democratic tool, the activists' description of how the direct action would unfold and the police response.

The Different Positions in the News Coverage

The analysis shows that the perspective '*Civil disobedience as a form of action*' was undoubtedly the dominant perspective in the debate. At the same time, this perspective was also the one that climate activists hoped would play the smallest possible role in the debate in the mass media.[8] This perspective, although especially dominant just before the Shut It Down action, and in the news coverage on the day of the action, was also present after it ended. This perspective comprised discussions and disagreements between numerous different actors, who challenged and contested each other's ideas and points of view, except on the day of the action, which is the focus of the next paragraph.[9]

The activists and Vattenfall were the clear opponents in the public debate in the perspective '*Vattenfall's use of coal*'. This perspective had substantial space in the Shut It Down debate in the months leading up to the direct action. The role of this perspective sharply declined in the news coverage in the days immediately prior to and during the day of the action.[10]

---

8   Based on insights from my observational study.

9   (Arbejderen 2009b, Arzrouni 2009, Christensen 2009, DR Deadline 2009a, Dr.dk 2009c, Fyhrie 2009, Jacobsen 2009, Modkraft.dk 2009e, 2009f, 2009g, Newspaq 2009, Nyboe 2009, P3nyhederne 2009, P4 Radioavisen 2009, Ritzau 2009a, Schmidt 2009, Stampe 2009, Voller 2009).

10   (Arbejderen 2009a, 2009c, Borking 2009a, 2009b, Damgård 2009a, Dr.dk 2009a, Grønbech 2009, Nyboe 2009, Sparevohn 2009, Voller 2009).

Lacking an opponent

Dean (2008) emphasises that having opponents who willingly engage and respond to the critical messages voiced is necessary if a debate is to have the possibility of becoming politically challenging. If the opinions stated do not receive any response then they will only be circulated and remain uncontested. Contestation is necessary condition for a topic to have the possibility of becoming constituted in the first place as a publically debated issue (Dean 2008, 101–3). The circulation of critical viewpoints that did not become contested was precisely the case for the two articulated perspectives – *'The climate and energy policies of the Danish government'* and *'The market and the consumers' role in the climate issue'*. These two perspectives lacked engaging opponents that challenged or disputed their arguments in the mass media. Although able to voice their critique and clearly name the government as their opponent in the mass media,[11] the media never sought a response from any one representing the government's stances. Furthermore, journalists did not seek comments on the issues raised in the news from experts or ordinary Danish citizens. Due to this lack of opponents and others relating or responding directly to the activists' critical messages covered in the news, these two perspectives were highly marginalised.

When news coverage turns into sports journalism

The *'Civil disobedience as a form of action'* perspective dominated news coverage extensively on the day of the Shut It Down event, turning coverage into what can be described as sports journalism, where the focus shifted from the topics raised to the action itself as a spectacle. The news reports from the day can be described metaphorically as a broadcast of a football match between two teams – the police and the climate activists. The journalists focused on the teams' tactics, their uniforms, and the team spirit just before the action kicked off and during the event. The journalists acted as good sports commentators providing their audience, which was not present at the power station, continuous updates on the standoff between the activists and the police. Ongoing updates from the online sites of major Danish newspapers and, in particular, the Danish news agency Ritzau,[12] functioned as uninterrupted sports-style commentary, providing instant updates on the action at the event as a spectacle. Following the traditions of good sports coverage, the two captains (spokespeople for both the police and the activists) were interviewed

---

11   (Borking 2009b, Damgård 2009a, Modkraft 2009c, 2009d, 2009h, Nyboe 2009, Urban Øst 2009, Voller 2009)

12   Ritzau's stories were broadcast largely unedited to almost every national, regional and local newspaper in Denmark, online or offline. Ritzau's coverage was, for the most part, the only source regional and local newspapers had on the Shut It Down action.

on how they thought the match had played out, after the action had ended.[13] The fact that the news journalism around the event turned into sports journalism in the mass media reduced critical points of the three other articulated perspectives, '*The climate and energy policies of the Danish government*', '*The market and the consumers' role in the climate issue*', and '*Vattenfall's use of coal*', to brief statements, banners and headlines. On the day of the action, when news coverage was at its most intense in terms of the entire public debate, there was no discussion of, or contesting arguments about, these other three critical perspectives raised by the activists. Furthermore, the discussion of civil disobedience as a legitimate democratic tool that caused the dissemination of dissenting ideas at other times during the debate, which is covered by the perspective '*Civil disobedience as a form of action*', was not present in press coverage either. News coverage on the actual day of the Shut It Down action remained limited to broadcasting the event as a spectacle to the public.

An Opponent

Vattenfall stood as a clear opponent to the climate activists in the articulated perspective, '*Vattenfall's use of coal*' in the debate. Criticised by activists for their use of coal at the company's Amager Power Station, Vattenfall could not avoid responding to the censure directly aimed at them in the mass media. In their response, Vattenfall tried to alter how they were being discursively framed as an unmistakable opponent to the activists' stance, a pattern of crisis communication that follows what Zizek has described as the dominant manner in which political issues are addressed today, which he calls post-politics. Post-politics portrays liberal democracies as power-neutral, spaces free of conflict in which the existence of real conflicts of interest between different actors in society are neglected. Instead, politics are reduced to looking for *solutions that work*. Thus a post-politics agenda reduces conflicts of interest to the need for finding the most efficient administrative and technical solutions, often through expert management. This way of articulating real political conflicts ignores the existence of genuine political differences between different parts in a society and thus makes the articulations of political conflict, which is a necessary condition for the possibility of social and political change, impossible (Zizek 2008: 34, 2007, 1999: 198–205).

The director of communications at Vattenfall tried to steer the corporation away from having the position of being the activists' opponent in the media by emphasising that they all shared the same goal regarding the use of coal for energy production: "We agree with the activists that the consumption of coal needs to be reduced in the long term (…)" (*my translation*, Borking 2009a).

---

13   (Frøslev 2009, Fyns Stiftstidende 2009, Jp.dk 2009, Nørmark 2009, Ritzau 2009c, 2009d, 2009e, Politiken.dk 2009, Schultz 2009)

Vattenfall avoided articulating a conflict of interest between the point of view of the activists and their own perspective when addressing the critique of the activists directly, and pointed out possible solutions to how coal could be phased out. Instead, Vattenfall took a post-political perspective and turned the critique into an issue about technological, financial and administrative problems that needed to be overcome on the road to handling climate change (Sparrevohn 2009, Borking 2009b).

The post-politics tendencies were clearly present when the ongoing debate between Vattenfall and climate activists focused on the role of Carbon Capture and Storage (CSS). Vattenfall described CCS as a necessary solution to climate change that had encountered technological and financial obstacles needing a resolution, before realisation as an important solution to climate change. Citing the financial crisis that led to a postponement of the company's CCS pilot projects, they also emphasised that the technology still needed to be fully developed (Sparrevohn 2009, Borking 2009b). The climate activists, on the other hand, focused on the consequences of prioritising CCS. They wanted CCS be abandoned, arguing that CSS would lead to a need for building larger power plants in Denmark and hinder a quick phasing out of coal. The issue highlights a conflict of interest between the activists and Vattenfall (Nyboe 2009). Furthermore, there has been strong local resistance to Vattenfall's CCS pilot project in the northern part of Denmark, which indicates that the power-neutral space free of conflicts of post-politics does not exist (Røttbøl and Nielsen 2009).

Vattenfall and the activists were opponents in the debate concerning the different political solutions to the use of coal. Consequently, it is not just a matter of agreeing on solutions that simply work: they do not exist. Completely abandoning coal as quickly as possible, compared to scaling down on the use of coal and relying on CCS to mitigate CO2 emissions, are two entirely different political takes on the relationship between coal and climate change. Vattenfall and the activists' different positions emphasise that the problem of climate change is a political issue that produces numerous political conflicts, which cannot be addressed and solved by relying on technological and administrative initiatives. The post-political articulation used by Vattenfall in the debate with the activists in the mass media stood in the way of politicising the activists' perspective, and thus the development of a debate with the potential to produce political challenges.

Critique and Neoliberalism

Throughout the debate, the activists' argumentation in the four perspectives articulated above focused chiefly on two specific actors – Vattenfall and the Danish government. With only a few exceptions, the climate activists generally failed to manage to link their critique to broader political demands of a more systemic nature in the public debate, in the form of demands for more structural or institutional changes to Danish climate politics.[14] This connection is crucial in order for the debate to become politically challenging, because topics and statements of opinions are only politicised if they are linked to broader social or political demands, as pointed out by Dean (2008).

The late Foucault's understanding of the term 'critique' also points to the necessity of a critique that articulates broader political demands if it is to be possible to politicise critical statements that are focused on specific actors. Foucault describes critique as a critical attitude that historically emerged as part of the Enlightenment, which also simultaneously produced a governmentalisation of Western societies, in which 'to govern' became the fundamental societal principle. A critical attitude is not connected to self-criticism, but exists in relationship to society and thus exists through, and grows out of, the relationships to the fundamental societal principle: to govern. Critique is thus not a question of 'how not to be governed', but a question of 'how not to be governed *like that*' (Foucault 2007). Foucauldian critique is about questioning or demanding changes in the dominant way of governing. It is an attempt to demand another way of governing and thus an effort to show that things could be another way; it is possible to govern differently. In the case of the climate activists their critical statements must go further than just criticism of specific actors and strive to question and challenge the dominant ways of governing in the field of Danish climate politics. Foucault characterises the practice of critique in the following way:

> I will say that critique is the movement by which the subject gives himself the right to question truth on its effects of power and question power on its discourses of truth. Well, then!: critique will be the art of voluntary insubordination, that of reflected intractability. Critique would essentially insure the desubjugation of the subject in the context of what we could call, in a word, the politics of truth (2007: 47).

The practice of critique must be understood as a way to challenge dominant power/knowledge relationships, which manifest themselves as politics of truth that provide the normative ground for policies. In the case of the Shut It Down debate, critique from activists needed to have focus on challenging the politics of truth that deliver the normative foundation for the dominant policies present in the political field of climate change. Consequently, activists' practice of critique must be about articulating perspectives, rationales, and ideas in the debate that challenge the widely accepted politics of truth which deliver the legitimising basis for Vattenfall and the Danish government perspectives on the use of coal.

As noted earlier, the activists nearly did not manage to articulate demands of a more systemic nature during the debate. A closer look at the few periods in the debate when the activists actually *did* manage to articulate more structural and

---

14   Note that voicing critique of this nature in the mass media is a communicative challenge for activists in today's mass media climate (Timms 2005).

institutional aspects in their critical messages is revealing. This makes it possible to discuss how an articulation of this kind, in which the outlined perspectives that challenge the normative foundation for the policies on the use of coal could have fostered a more politically challenging debate, if it had gained a stronger voice in the debate.

A critique with broader structural perspectives

The articulated perspective, '*The role of the market and consumers in the climate issue*', was the underlying normative ground for the critique articulated by the activists in the debate (Borking 2009b, Modkraft 2009c, 2009h, Nyboe 2009, Voller 2009). The activists were highly critical of the logic of the market, where people act as consumers, as the key to solving climate change. This criticism reflects some of the characteristics and criticisms numerous scholars have linked to neoliberalism (Chiapello and Boltanski 2005, Hesmondhalgh 2008, Lemke 2002, Lipschutz 2005, Oels 2005, Urry 2010, Willig 2009).

Before taking a closer look at the point in the debate when the activists most clearly formulated a criticism with broader structural perspectives, giving an account of the term neoliberalism is necessary. The account used here follows the perspective in Foucault's governmentality studies, where neoliberalism must be understood as a neoliberal governmentality that produces a specific relationship between the state and the market (Foucault 2008, 2009). This starting point enables an understanding of how the basis for a wider articulated critique of a more systemic nature could have been had the activists been able to get this kind of critique across in the news coverage more strongly. The historical-philosophical analyses of the concept of government that governmentality studies provide offer many important insights, a few of which I will focus on below.

With the rise of the modern state, 'to govern' became the fundamental principal for society; the state governing the whole population to secure the prosperity and welfare of all inhabitants; this is still the case today. Foucault points out that this development led to the emergence of political economy in which politics and economics became interlinked in governing. From that point on, understanding economics, without looking at its connection to political action, has been impossible. The market and the politics of the state are therefore interconnected. Foucault's analysis of neoliberalism as a specific neoliberal governmentality shows that the purpose of the neoliberal art of government is to ensure a successful development of the market is supported. The state must *govern for the market* by actively creating conditions that enhance the actions of the individuals that underpin the success of the market and by creating legislation that supports market success. When the state governs for the market, sustained economic growth becomes the parameter of success for the state's ability to ensure the prosperity and welfare of the population (Foucault 2009: 95–240). Foucault's account of neoliberalism as a neoliberal governmentality makes it clear that neoliberalism should not be understood in terms of the role of the state has been downgraded to ensure the market as the dominant logic in society. Instead, it is a governmentality that fosters an actively governing state, where the goal and guiding principle of the art of government is to assure that the market is as successful as possible.

Following Foucault's understanding of neoliberalism as a governmentality, German sociologist Thomas Lemke highlights that neoliberalism must be seen as a condition that produces a certain rationality, which functions as the justification and guiding principle for the dominant way society is governed today (Lemke 2001, 2002: 54–60). Keeping Foucault's understanding of critique in mind, a neoliberal rationality will be manifested in numerous politics of truth that provide the normative ground for the dominant policies on the use of coal in Denmark. Foucault (1991b: 79) and Lemke (2002: 54–61) stress that any way of governing has a specific rationality. Neoliberal governmentality must be seen as one form of rationality among others, each of which can be justified as the guiding principle for how to govern. In short, neoliberal policies are grounded in a specific rationality and politics of truth and must be understood as part of a specified neoliberal governmentality, where the guiding principles for the policies are to ensure a way of governing that governs for the market with the goal to produce the most successful market as possible.

I will, from the Foucauldian perspective on neoliberalism, expand on the point in the debate where the activists managed to clearly articulate a critique with broader structural perspectives. This enables an understanding of what the basis for a wider articulated critique of more systemic nature could have been, had it gained more prominence in the news.

With the EU's liberalisation of the electricity market, it has been possible for companies like Vattenfall to continue to smash the earth because it is legal to choose the cheapest form of production rather than the most appropriate. This shows that market forces cannot stop a climate catastrophe, because growth and profit will always be valued more than the environment. A profit-seeking corporation like Vattenfall will always be a part of the problem. (...) We have been given empty words from both the government and the power companies that everybody agrees and is working towards the same goal. The obvious conflict that exists between the interests of profit and the environment are downplayed and tucked away behind smart words. It is important for us to say no to this spin and maintain that there is an irresolvable conflict between our wishes for a world without coal and the market's need for growth. (...) Energy policies must not be left to the market but should be based on democratic decisions, which prioritise the common good over the interests of profit (*my translation*, Nyboe 2009).

The climate activists presented an unambiguous contrast between market forces and profit on the one hand and environmental issues and solving climate change on the other. This is the case for both government policies and Vattenfall's actions. The demarcation the activists make is especially clear towards the government, where the activists emphasise that energy policies have been relinquished to market forces. When it comes to Vattenfall, the activists initially point to the EU's liberalisation of the electricity market, a way of governing, as the reason for the company's continued use of coal. They then linked this claim to the articulated contrast between profit and the environment, where Vattenfall, according to the activists, will always focus on pursuing profit over the environment.

The activists lack the perspective that neoliberal governmentality governs for the market, where a successful market is seen as the condition that ensures the common good. Thus, energy policies reliant on the market logic follow a neoliberal political rationality, where economic growth and a well-functioning market are seen as the driving forces that can provide change regarding the use of coal and dealing with climate change. The activists can be right in pointing out that this way of governing, which legitimises the logic of the market as the fundamental source to solve climate change and phase out coal is problematic, but within neoliberal policies there is no perceived conflict between ensuring the welfare and prosperity of the population, and hence the common good, and a market that delivers profit. From a neoliberal viewpoint, the market's success provides prosperity; therefore it is not a way of governing that leave policies to the market at the expense of the common good and democratic decisions for the sake of profit, as the activists argue. Instead, neoliberal governmentality fosters an active way of governing that makes the logic of the market the crucial tool for, e.g. phasing out the use of coal and addressing global warming and the underlying principle of ensuring the common good.

To make a more politicised debate possible, the activists would have needed to abandon the idea that the government and Vattenfall are using rhetorical moves and spin to cover: "an irresolvable conflict between our wishes for a world without coal and the market's need for growth" (Nyboe 2009) as there are, per definition, no conflicts here in this neoliberal way of governing. The activists would instead have needed to engage the dominant climate and energy politics as based on a political rationality grounded in a neoliberal governmentality. This approach would have involved articulating a critique with systemic perspectives that questioned and challenged the political focus on and the active governing of the market as the way to end the use of coal and deal with climate change. Such a critique would have required formulating criticism that did not just focus on a conflict between the environment on the one side and profit and economic growth on the other. This critique would have also required criticism with the ability to formulate systemic demands for a different relationship between the market and the state, where the guiding principle for the way of governing was not merely the success of the market. This approach would involve articulating demands for the formulation of climate and energy policies that address the market as problematic for climate change and for eliminating the use of coal and, in this sense, demanding another way of governing not based on the politics of truth and the rationality of a neoliberal governmentality. Such a critique would include the formulation of empirical demands for concrete political actions to address and centre on altering the focus of climate and energy policies on securing the success of the market.

Conclusions

The most political challenging part of the Shut It Down debate sparked in the mass media was present in the articulated perspective '*Civil disobedience as a form of action*', in which several different actors responded to each others' perspectives. However, this perspective, which turned out to be the most dominant one in the debate, was not a part of the issues raised by the activists.

This chapter has hopefully made it clear that the Shut It Down debate faced quite a few obstacles preventing it from becoming a politically challenging debate. First, the critique from the activists represented in '*The climate and energy policies of the Danish government*' and '*The role of the market and consumers in the climate issue*' perspectives became marginalised in the debate by a lack of opponents who responded to the criticism circulated. This marginalisation was supported by the news coverage, which did not have experts, representatives of the government's viewpoint, or ordinary citizens relate or respond to the criticism represented by the two perspectives. Furthermore, news coverage on the day of the protest action turned into sports journalism, with the focus shifting from the topics raised to the action as a spectacle. This shift reduced the critical aspects in the three other articulated perspectives – apart from '*Civil disobedience as a form of action*' – to short statements, banners, and headlines.[15]

The second of these obstacles was Vattenfall as an opponent, since the company tried to manoeuvre out of its role as an opponent of the critique, by not recognising the conflicts of interests between themselves and the activists. Instead, the company tried to turn the criticism raised into a post-political discussion in which the activists and Vattenfall had common interests, highlighting the issue as simply a question of overcoming technological, financial and administrative challenges to deliver solutions to cope with climate change and end the use of coal. The lack of clear articulation of different interests made the debate less politically challenging.

Finally, the activists generally did not manage to articulate a critique with broader structural perspectives. When they managed to formulate this kind of criticism, they articulated a critique of profit being pursued at the expense of the environment and the climate. This stood in the way of the possibility for a critique that would question and challenge the neoliberal way of governing, where the role of climate and energy policies from the state is to ensure the success of the market as the way to handle global warming and end the use of coal. The activists' sole focus on the conflict between market forces and the environment lacked a clear formulation of a critique addressing the pre-eminence of neoliberal, pro-market politics as the solution to climate change and ending the use of coal. Clearly, if the activists wished to have

---

15   Note that the pattern the news coverage followed of turning into sports journalism, where the focus was on the action as a spectacle, was repeated in the Danish news coverage of the largest direct action carried out during COP15, the United Nations Climate Change Conference in Copenhagen in December 2009 (Astrup et al. 2009, Haslund 2009, Mølgaard 2009, Rydzy 2009b).

greater success, they needed to articulate an alternative view of the relationship between the state and the market in climate and energy policies.

References

Theoretical sources

Almlund, P. 2007. Miljøkommunikation i virksomheder – praksis i kontekstens blinde plet. Roskilde: Roskilde University.

Boltanski, L. and Chiapello, E. 2005. The New Spirit of Capitalism. London: Verso.

Bourdieu P. 2001. Modild – for en social bevægelse i Europa. Copenhagen: Hans Reitzel Publishers.

Danish Commission on Climate Change Policy 2010.Green Energy – the road to a Danish energy system without fossil fuel, Copenhagen: Ministry of Climate and Energy.

Dean, J. 2008. Communicative Capitalism: Circulation and the Foreclosure of Politics, in Digital Media and Democracy: Tactics in Hard Times, edited by M. Boler. Cambridge, MA: The MIT Press, 101–21.

de Jong, W., Shaw, M. and Stammers, N. 2005. Introduction, in Global Activism, Global Media, edited by W. de Jong et al. London: Pluto Press, 1–14.

Flyvbjerg, B. 1991. Rationalitet og Magt, Bind 1 – Det konkretes videnskab. Aarhus: AkademiskForlag.

Foucault, M. 2007. What Is Critique, in The Politics of Truth. Los Angeles: Semiotext, 41–81.

Foucault, M. 2008.Stat, Territorium og Befolkning. Copenhagen: Hans Reitzel Publishers.

Foucault, M. 2009. Biopolitikkens fødsel. Copenhagen: Hans Reitzel Publishers.

Gaber, I. and Willson, A.W. 2005. Dying for Diamonds: The Mainstream Media and NGOs: A Case Study of Action Aid, in Global Activism, Global Media, edited by W. de Jong et al. London: Pluto Press, 95–109.

Haahr, J.H. 2000. Offentligheden: fornuftskilde eller magtarena? Om Jürgen Habermas' offentlighedsforståelse og Foucaults kritik. Aarhus: Danish School of Media and Journalism.

Hesmondhalgh, D. 2008. Neoliberalism, Imperialism and the Media, in The Media and Social Theory, edited by D. Hesmondhalgh et al. London: Routledge, 95–111.

Højbjerg, H. 2005. Hermeneutik, in Videnskabsteori i samfundsvidenskaberne. Roskilde: Roskilde University Press, 309–47.

Kristiansen, S. and Krogstrup, H.K. 1999. Deltagende observation. Introdukton til en Forskningsmetodik. Copenhagen: Hans Reitzel Publishers.

Lemke, T. 2001. The birth of bio-politics: Michel Foucault's lecture at the Collége de France on neo-liberal governmentality. Economy & Society, 30(2),190–207.

Lemke, T. 2002. Foucault, governmentality, and critique. Rethinking Marxism, , 14(3), 49–64.

Lipschutz, R.D. 2005. Networks of knowledge and practice: Global civil society and global communications, in Global Activism, Global Media, edited by W. de Jong et al. London: Pluto Press, 17–33.

Oels, A. 2005. Rendering Climate Change Governable: From Biopower to Advanced Liberal Government? Journal of Environmental Policy &Planning, 7(3), 185–207.

Schultz, I. 2006. Bag Nyhederne. Frederiksberg: Samfundslitteratur.

Timms, D. 2005. The World Development Movement: Access and representation of globalisation: Activism in the mainstream press, in Global Activism, Global Media, edited by W. de Jong et al. London: Pluto Press, 125–32.

Urry, J. 2010. Consuming the Planet to Excess. Theory, Culture & Society, 27(2–3), 191–212.

Willig, R. 2009.Umyndiggørelse. Et essay om kritikkens infrastruktur. Copenhagen: Hans Reitzel Publishers.

Zizek, S.1999. The Ticklish Subject. London: Verso.

Zizek, S. 2007. Censorship Today: Violence, or Ecology as New Opium for the Masses. Lacan dot com, http://www.lacan.com/zizecology1.htm.

Zizek, S. 2008. Violence. London: Profile Books.

Empirical sources

Arbejderen 2009a. Vattenfall vil ikke i dialog. *Arbejderen*, 19 September.

Arbejderen 2009b. Luk ned for kulkraft! *Arbejderen*, 24 September.

Arbejderen 2009c. Aktivister lukker Amagerværket, *Arbejderen*, 29 August.

Arzrouni, C. 2009. Debat: Mod strømmen: Farlige klimaalliancer. *Jyllands-Posten*, 26 October.

Astrup, S. 2009a.Klimaaktivister: Vi vil afbryde topmødet. *Politiken*, 8 December.

Astrup, S. 2009b. Udsigt til weekend med aktioner. *Politiken*, 9 December.

Astrup, S., Hvilsom F., Søndergaard, B., Berndt, T. and Bennetsen, D. 2009. Demonstrationen onsdag minut for minut. *Politiken*, 16 December.

Batchelor, O. 2009. Aktivister: Lømmelpakke får flere på gaden. *Newspaq*, 16 September.

Borking, L. 2009a.Miljøkamp: Aktivister vil lukke kraftværk på Amager. *Information*, 18 August.

Borking, L. 2009b. Aktivister vil forsøge at lukke kulværk. *Information*, 25 September.

Borking, L. 2009c. Frysende pingviner i aktion mod kul. *Information*, 28 September.

Christensen, V. 2009. Debat: Ligegyldigheder på tv. *Jyllands-Posten*, 3 October.

Damgård, T. 2009a. Aktivister vil lukke kraftværker. *Berlingske Tidende*, 3 September.

Damgård, T. 2009b. De pæne gamle, de unge aktive og de helt sorte. *Berlingske Tidende*, 3 September.

Dr.dk 2009a. Amagerværket: Vi er enige. *Dr.dk*, 18 August.

Dr.dk 2009b. Amagerværket vil stadig blive besat. *Dr.dk*, 18 September.

Dr.dk 2009c. Sådan vil aktivisterne lukke Amagerværket. *Dr.dk*, 21 September.

Dr.dk 2009d. Landmænd og miljøaktivister vil kæmpe sammen. *Dr.dk*, 29 September.

DR2 Deadline 2009a. Der skal slås hårdt ned på demonstranter. *DR2 Deadline 17:00*, 15 September.

DR2 Deadline 2009. ShutIt Down i alliance med halmleverandører. *DR2 Deadline 17:00*, 29 September.

Eising, J. 2009. Razzia mod aktivister. *Berlingske Tidende*, 09 December.

Ekstra Bladet 2009. 1–0 til politiet. *Ekstra Bladet*, 14 December.

Frøslev, L. 2009. Masseanholdelser ved klimademonstration. *Berlingske Tidende*, 26 September.

Fyhrie, E. 2009. Fra ugens netdebat: Den nødvendige civil ulydighed. *Information*, 8 October.

Fyns Stiftstidende 2009. Politiet stoppede klimakamp. *Fyns Stiftstidende*, 27 September.

Grønbech, J. 2009. Kulaktivister bekriger kraftværk. *Newspaq*, 18 September.

Hansen, J. 2009. Civil ulydighed ved afgrundens rand. *Information*, 17 August.

Hvilsom, F. 2009. Østre Landsret advarer mod lømmelpakke. *Politiken*, 11 September.

Jacobsen, F. 2009. Debat: Aktivisterne bør betale. *Berlingske Tidende*, 5 October.

Jp.dk 2009. Aktivister afvist ved Amagerværket. *Jyllands-Posten*, 26 September.

Krause på Tværs 2009. Man skal ikke være bange for os. Interview med Tannie Nyboe, talskvinde fra klimaaktivistnetværket ClimateJustice Action. *Krause på Tværs DR P1*, 28 October.

Liversage, T. 2009. Civil ulydighed er en pligt. *Information*, 29 September.

Modkraft.dk 2009a. Vattenfall til aktivister: Vi vil ikke i dialog. *Modkraft.dk*, 17 September.

Modkraft.dk 2009b. Aktivister til politiet: Vi er IKKE en trussel imod almindelige samfærdsmidler. *Modkraft.dk*, 24 September.

Modkraft.dk 2009c. Aktivister vil lukke kulkraftværk. *Modkraft.dk*, 2 June.

Modkraft.dk 2009d. Aktivister lukker for Amagerværket 26. September. *Modkraft. dk*, 18 August.

Modkraft.dk 2009e. Klima-Aktivister: Lømmel-pakke vil optrappe konfrontationer. *Modkraft.dk*, 15 September.

Modkraft.dk 2009f. Stop dansk kulkraft – Luk værket ! *Modkraft.dk*, 26 September.

Modkraft.dk 2009g. Civil ulydighed mod kul på Amagerværket kan godt forsvares. *Modkraft.dk*, 26 September.

Modkraft.dk 2009h. Klimaaktivister mobiliserer på Roskilde. *Modkraft.dk*, 4 July.

Mølgaard, M. 2009.Klimaaktivister opgiver at trænge ind i Bella. *Politiken*, 16 December.

Nyboe, T. 2009. Kommentar: Klodens fremtid bør ikke sættes til salg. *Information*, 8 September.

Nørmark, T. 2009. Lukket ned. *BT*, 27 September.

P3 Nyhederne 2009. Tidobling af bøder. *DR P3 08:00*, 15 September.

P4 Radioavisen 2009. Bøder afskrækker topmødeaktivister. *DR P4 09:00*, 15 September.

Politiken.dk 2009. Aktivister spærret inde på Amagerværket. *Politiken*, 26 September.

Ritzau 2009a. Aktivister vil lukke kraftværker. *Ritzau*, 3 September.

Ritzau 2009b. Amagerværket roser politiets indsats. *Ritzau*, 26 September.

Ritzau 2009c. Miljøaktivister stopper demo på Amager. *Ritzau*, 26 September.

Ritzau 2009d. Aktivister på vej til Amagerværket. *Ritzau*, 26 September.

Ritzau 2009e. Klar til endnu et kraftværk. *Ritzau*, 29 September.

Rydzy, M.S. 2009a. Klimaaktivister: Vi har stor opbakning. *Berlingske Tidende*, 8 December.

Rydzy, M.S. 2009b. Interview: I 15 år er der intet sket. *Berlingske Tidende*, 9 December.

Rydzy, M.S. and Haslund, E.A. 2009. Aktivister har forladt Bella Center. *Berlingske Tidende*, 16 December.

Røttbøll, E. and Nielsen, J.S. 2009. Kampen for klimaet går under jorden. *Information*, 15 June.

Schmidt, R.K. 2009. Både aktivister og lovgivere truer brugen af civil ulydighed. *Information*, 29 October.

Schultz, T. 2009. 150 aktivister anholdt. *Ekstra Bladet*, 27 September.

Sparrevohn, M.R. 2009. Kommentar: Vi arbejder hårdt for at nedbringe kulforbruget. *Information*, 15 September.

Stampe, C. 2009. Klimastraf: Eksperter: Højere straffe til klimaaktivister nytteløse. *Information*, 16 September.

Urban Øst 2009. 1... 2... Hurtige 3. *Urban* Øst, 24 September.

Voller, L. 2009. Aktivisme: Hvis politikere ikke lukker kulkraftværkerne, så må vi jo. *Information*, 18 July.

Chapter 4.5

Negotiating and Communicating Climate

Pernille Almlund

The United Nations Framework Convention on Climate Change (UNFCCC) meeting, popularly known as the 15th Conference of the Parties (COP15), was held in December 2009 in Copenhagen, Denmark. Up to and during COP15, many politicians, researchers and media representatives described the summit as a critical step in developing a global response to the threat of climate change caused by human activity. Held annually, COP meetings are and continue to be a forum for the on-going process of international negotiation and political communication on climate change. The COP15 goal was to agree on a new climate protocol to replace the Kyoto Protocol adopted on 11 December 1997 that entered into force on 16 February 2005. This objective was not reached at COP15, and instead of a legally binding agreement, the UN launched a political one, the Copenhagen Accord.

Denmark campaigned hard to host COP15, resulting in high expectations among not only Danish politicians in particular but also among politicians worldwide. Consequently, the importance of the summit led to an exceedingly strong political and media-influenced agenda on climate change, nationally and internationally, that covered the entire gamut from climate meetings to climate performance.

In this chapter I focus on the political agenda, negotiations and communication on climate change that occurred nationally in Denmark, where hosting COP meant special issues involving the climate were brought up. For example powerful efforts were made to brand the country as a climate frontrunner both politically and among large enterprises. This chapter addresses the question: *What were political communication and the negotiations on the climate like in the Danish Parliament around the time COP15 was held in December 2009 and what are they like today?*

In 1986 Niklas Luhmann wrote, Ökologissche *Kommunikation: Kann die moderne Gesellschaft sich auf* ökologische *Gefährdungen einstellen*, which was published in English in 1989 as *Ecological Communication*. Based on Luhmann's general system theory, the book underlines how modern society is a differentiated society where unity does not exist and society is separated into social systems such as the economic system, the political system, the religious system and the education system. The specific nature of these systems in Luhmann's theory is autopoietic, which means they are self-producing, thus underlining the operational closure of social systems (Luhmann 2002; 1998; 1992). In this sense every social system operates within its own code, programmes and logic and is only able to handle disturbances from its environment and other surrounding systems within its own code. A highly differentiated society means that having a common understanding of and jointly solving ecological problems is impossible because each social system's perception and resonance of the ecological issues at hand are different (Luhmann 1989). Luhmann describes how a lack of acceptance concerning the differentiation of society creates a barrier when it comes to solving environmental issues:

> Our aim was to work out how society reacts to environmental problems, not how it ought to or has to react if it wants to improve its relation with the environment. Prescriptions of this sort are not hard to supply. All that is necessary is to consume fewer resources, burn off less waste gas in the air, produce fewer children. But whoever puts the problem this way does not reckon with society, or else interprets society like an actor who needs instruction and exhortation (....) (Luhmann 1989:133)

Luhmann's idea is that the probability of communication about ecological issues entering the political system is high due to the political system's internal operations and methods, namely the production of collectively binding decisions. These decisions should be seen as internal political affairs and rest on the internal logic of the political system, which is to be sensitive to power and, consequently, voters. The result is that collectively binding decisions have no direct environmental impact and will merely have an internal social effect. Moreover, the decisions will only be able to influence other systems and not regulate them. In this regard, Luhmann explains the necessity of completing further empirical research on the political system's perceptions as well as on communication about environmental issues (Luhmann 1989). This chapter thus responds to Luhmann's call for additional empirical research and addresses the climate as the specific environmental issue.

The Theories, Methods and Empirical Elements

In order to answer the research question posed in this chapter a study was done of the negotiations that occurred concerning the laws that were established and enacted, as well as the enquiries that took place and motions that were proposed on the climate in the Danish Parliament from 1 November 2009 to 30 March 2010. The data comprises transcriptions routinely taken in the Danish Parliament and published in *Folketingets Tidende* (*News from the Danish Parliament*). An examination of the transcripts shows not only how political negotiations on the climate took place, but also how the political system understands the climate issue. Face-to-face, in-depth interviews were also conducted with the climate spokespeople for all of the parties represented in the Danish Parliament. These interviews were carried out from February 2010 to August 2010.[1]

---

1   The knowledge and perspective of the different Members of Parliament interviewed are likely to have been influenced by this relatively long time span, which was unavoidable

The semantic and form analysis employed in this study are inspired by Luhmann. The analysis focuses on the thematic concept of climate communication praxis based on the system theory of Luhmann and the praxis theory of Pierre Bourdieu. The aim of this chapter is to show what the communication praxis among politicians was on the climate as well as describe the dominant form of communication on the climate in the political system. With respect to the communication praxis of politicians, determining whether or not there is a specifically political way of "talking about the climate" is of interest. An analysis was carried out, for example as to whether the level of agreement among politicians was as high as the newspapers indicated in the months up to and during COP15. In addition, how the ideology, scientific knowledge, media and voter opinion influence the semantics and understanding of the politicians is also examined.

Theoretical Inspiration

Using Luhmann's theory of social systems the analysis draws on his concepts of communication and observation, as well as his understanding of society as differentiated and his thoughts on ecological communication. Systems are in the system theory of Luhmann understood as the unity of the difference between system and environment. Thus no system can exist without an environment and all other systems will be part of the specific system's environment.

As stated, the understanding of society as differentiated establishes the perception, communication and action concerning particular issues such as ecological problems or climate change as system specific. Every functional system, e.g. scientific, political and economic, operates within its own specific binary code. Social systems identify themselves by their binary codes, distinguishing themselves from their environments by the specificity of each individual code. Furthermore the binary code is a strictly internal structure (Luhmann 1992). For the political system, the binary code is power in opposition to not power (power/ not power); for the scientific system, it is truth in opposition to false (truth/false); and in the economic system, it is pay in opposition to not pay (pay/not pay). Consequently, the communication of different systems differs and the different systems understand issues, as per the example of climate change, in exceptionally different ways. In general the political system communicates about climate change as a means for gaining voters and power, while the scientific system communicates about climate change to achieve the truth and the economic system communicates about it to achieve the best financial solution. Communication on climate change in each specific system will of course be different depending on e.g. the political opinions of politicians, the theoretical choices of researchers and the economic understanding of economists. Thus the systems are operationally closed but still cognitively open, which means that they are able, in Luhmannian terms, to irritate

due the MPs' busy schedules.

other systems and be irritated by other systems and their environment in general. The handling of these irritations is system specific. If the different systems then act and communicate about climate change, their perception and what Luhmann calls resonance of this will be system specific. Furthermore the systems will also only be disturbed and show resonance if climate change somehow fits into the specific system's way of perceiving and communicating (Luhmann 2000; 1989).

In system theory, systems are constituted by communication, while the acknowledgement and perception of the systems and society happen through observation of communication. In Luhmanian terms, to observe is to draw a distinction between what is marked and what is unmarked. This means that when something is indicated or marked, something else will be excluded or unmarked. In that sense, to observe is to draw a distinction and the observation is the unit of the distinction. Even though the marked side of the distinction is thus inseparably linked with the unmarked side, the observer only observes the marked side and what is included in the observation and is unable to see both what is excluded and the unit of the distinction when the observation takes place. Every observation is in that sense based on a blind spot (Kneer and Nassehi 1997; Luhmann 1997; Thyssen 1997).

In time, it is possible for the observer to cross the boarder of distinction and see what was unmarked in the first observation. Luhmann calls this kind of observation a second order observation. First order observation is the observation and thus an indication of something in the world, while second order observation is the observation of observations. First and second order observation share a common state and both produce blind spots in their operation of observation. The only difference is that a first order observation is unaware of the blind spot, which is not the case for a second order observation, where it is well known (Luhmann 1995; Åkerstrøm Andersen, 1999).

The concept of first and second order observation also provides an opportunity for questioning common sense. A first order observation takes the observation for granted, in contrast to a second order observation, which observes what is taken for granted, thereby revealing what can be questioned (Kneer and Nassehi 1997; Luhmann 1997; Åkerstrøm Andersen 1999). Furthermore the search for similarity and unity in these exceedingly different observations can help us to identify the dominant form of communication in the system being researched. The dominant form of communication is the form of communication all of the communication in a particular system refers to (Åkerstrøm Andersen 1999). Identifying the dominant form of communication opens up the opportunity to more profoundly question what is taken for granted but not directly observable.

In keeping with Luhmann's concepts, to observe observations is to observe communication. Communication constitutes social systems and includes speaking, writing and gesticulating. As a result communication should, as mentioned, be observed when knowing what is going on in society is necessary. According to Luhmann, communication is thus neither communicational actions nor a transmission from a sender to a receiver. Luhmann defines communication as a three-pronged process of selection comprising the selection of information, the selection of an utterance and the selection of understanding. Communication does not become a reality until all three selections have taken place. The selection of the third element, understanding, does not refer to what a receiver receives but should be understood as the further communication by the connection to the message. From this it follows that only communication is able to communicate and communication is not communication between individuals (Luhmann 2000).

Consequently Luhmann's concept of communication is a non-individual concept of communication, which dissociates itself from the idea of intersubjectivity. As the constituter of social systems covering speaking, writing and gesticulation and as available for observation, this concept of communication includes all types of communication and becomes a concept that is more analytical than a traditional concept of communication. However, such a broad concept of communication requires a thematic delimitation of the communication intended for observation and is in this particular study communication on climate issues, i.e. climate communication in the political system.

In addition to an epistemological approach, this study is also based on the concept of practice as presented by Pierre Bourdieu in his theory of practice (1977). The concept of communication is not one commonly used by Bourdieu, although his latest book, *Science de la science réflexivité*, published posthumously, emphasises the importance of communication in the establishment of fields and communication is understood as the medium of symbolic capital, which should be seen as an important part of the organising factor of the power balance in the field (Bourdieu 2005 p. 62). The strong connection between field, habitus and capital in the praxis theory of Bourdieu indicates that communication is important for more than the establishment of fields. Communication also influences the constitution of habitus and capital and vice versa. This underlines the development of the concept of 'communicational practice' and should be seen as an important supplement to the concept of communication inspired by the system theory of Luhmann.

According to Bourdieu context is built upon the interaction of field, habitus and capital and thereby understood as e.g. experience, history, language and structures. Bourdieu sees context as a comprehensive concept and as based on both ontology[2] and epistemology (Bourdieu 1997; 1996; 1977). I draw on this understanding and see communication as a practice founded in the context. Inspired by Luhmann, I see communication as an excellent object of observation, but I also see communication as bounded by context as understood in Bourdieuian terms. Hence the observation of communication will also provide insight into the context. In this regard – following Bourdieu – it is important to point out that gaining full insight into the context is impossible as the theoretical practice will always only be an approximation of the practical practice (Callewaert 1997–1998).

The works of Luhmann and Bourdieu provide inspiration on different levels. Luhmann's concepts of communication and observation as well as his

---

2   Bourdieu views ontology not from a traditional perspective or as a search for the truth, but as a dynamic, and in some senses, relativistic concept.

thoughts on ecological communication are remarkably fruitful when applied in the exploration of the field of climate communication. Luhmann's exclusively epistemological focus is, however, at times limiting, which is why I draw on the ontological foundation put forward by Bourdieu. In the concrete analysis I apply only Luhmann's concepts, but within a Bourdieuan framework of understanding and use the concept communicational practice instead of solely communication. I am not exploring the ontological level, but centre on communicational practice understood as being dynamically ontologically founded. Consequently I utilise the concepts and analytical strategies of Luhmann, but I also understand and employ them based on this idea of dynamic ontology.[3]

Semantic Framework

As an observer of climate communication and in search of an answer to the proposed research question, I allowed the distinction between climate / not climate or what is marked as climate versus what is excluded as not climate, to guide me. The data produced; the interviews conducted with climate spokespeople in the Danish Parliament; and documents from *Folketingets Tidende* are what made the climate communication observable.

When exploring the variety of ways the Danish Parliament understands climate, it is important to note that climate should be understood as a concept and not as a fully defined unit. In Luhmanian terms a concept is given meaning by the context[4] and could be given meaning in numerous ways. From this perspective a concept is limited by its counter concept. Because of that the counter concept of climate in the political system was studied in the communication.

The observation of the understanding of climate issues is a *first order observation* of what is taken for granted as being in the world. Taking a specific understanding, for example about the climate, for granted, the meaning of the climate concept can often be a category and/or specific activities connected to a specific concept; or in other words it can be the way the climate concept takes shape. As a result the interviews and documents were analysed to categorise aspects involving the climate and for the type of activities mentioned as climate

---

3   For a more comprehensive explanation and account of the ontological understanding and the meta theoretical basis behind the combination of Luhman's system theory and Bourdieu's theory of practice see *Environmental communication in enterprises: Practice in the blind spot of context (Miljøkommunikation I virksomheder – praksis I konekstens blinde plet)* (Almlund 2007).

4   Luhmann defines context as the concrete context influencing communication and solely at the epistemological level. I define context as described above, but agree that concepts are given meaning by the context. Hence the constructivist inspiration behind this chapter remains intact.

activities. This approach revealed three specific types of climate communication or communicative practice taking place on the climate in the Danish Parliament:

1. Climate / counter concept
2. Categorising the climate
3. How the climate takes shape

These three types of climate communication are the communicative practice that emerged in the interviews and the documents and should be understood as the system's first order observations of climate. When it comes to a topic or concept like the climate in the political system many of the observations observed will be reflexive as politicians are fully aware of how their political opinions are changeable. Thus the observation of the understanding of climate issues can, and probably will, also include *second order observation*.

Even if the political system and thereby politicians are reflexive, they communicate about the climate using political statements that their specific party takes for granted. Thus it follows that most of the observations appear to be of the first order. My observations and analysis examine what seems to be taken for granted in the political system when the communication is about the climate.[5]

Semantic Analysis of Climate Communication in the Danish Parliament

While examining the semantic understanding of climate in the political system, I selected text that shows how the politicians communicate what climate is in different ways. In the early stages of reading *Folketingets Tidende*, the diction and topics of the parties show that all of them, on one level, have apparently strikingly similar definitions of climate. All of the parties primarily take the facts surrounding climate change for granted in their political speeches and negotiations. Accordingly, the degree of concurrence between parties at this level is prominent on the following issues:[6]

Climate is security policy.
Climate is development aid for the Third World.
Climate is public information initiatives.
Climate is a multiplier of threats.

---

5   This distinction is mentioned to show that politicians do both, but is not expounded upon further as the aim of this analysis is not to show how and when the political system – represented by the politicians – carries out first and second order observation. Both types of observation can be the object of observation and show how climate is understood in the political system.

6   Not necessarily every politician has mentioned every one of these topics, but all of these topics are mentioned by more than one politician.

> Climate is the reduction of CO_2 emissions – mitigation.
> Climate is society adapting to climate changes – adaptation.

Disagreement of course exists when negotiations take place, but not with regard to what they find worth mentioning as topics or as definitions of climate. More precisely, during interviews, several politicians state that there is agreement about the goals, but not the means for reaching those goals. Disagreement arises about the means, but is often just a digression where the parties can fight a little about well-known, specific disagreements between parties, e.g. concerning issues such as the EU and the UN. The data indicates that these differences regarding the climate are not especially conspicuous.

During the interviews with climate spokespeople, in contrast to reading *Folketingets Tidende*, this concurrence between parties is not evident. As a result, based on my reading of the negotiation transcripts, the highly framed, rule-bound and controlled discussions taking place in the Danish Parliament can only partially reveal how the climate is understood there. Perhaps this type of framed and controlled political discussion leads to agreement or at least makes the occurrence of the same definitions and choice of topics more probable, or at least makes them appear more probable. Thus official negotiations in Parliament seem to be more a matter of performance than actual negotiations. Commenting on negotiations taking place in Parliament, one politician states:

> If you understand the technique and if you know how to read it, you can also learn who has which point of view from a debate. But if you have to describe in your own words the scope of the interaction, then it can be very different – this means that, with words you can force yourself to a very high level, but real action can be very far away from those words.

Maybe this is the reason why it is very difficult to observe the counter concept of the climate and the categorisation of climate during negotiations in Parliament. The way climate takes shape or the specific activities connected to the concept is expressed in all of the definitions mentioned above, but this is not enough to draw up a more comprehensive picture of climate communication and how the Danish Parliament understands climate.

Portraying a diversified picture of the existing understanding of climate in the political system requires a study of the interviews conducted and the written documents. A combined analysis shows that each party categorises the climate in a specific way; namely each party has its own explicit, and for the party, extraordinarily coherent *political storyline* on the climate. These stories provide a comprehensive foundation for the further analysis of how climate is understood and what the dominant form of communication in the climate communication of the Danish Parliament is.

Political Storylines

All of the parties in the Danish Parliament are represented in these *political storylines*. The left wing[7] parties are Red-Green Alliance, the Socialist People's Party, the Social Democrats, the Danish Social Liberal Party, the Conservative People's Party, Venstre (Left), Liberal Alliance and the Danish People's Party.[8]

During the interviews it was possible to identify the following specific foci as important to the political storylines: the climate as an important political issue; science; third-world and supplementary benefits; the press and different medias; technology, organisations; energy and security policy; goals and means; and time. Nearly all of these foci appear in every storyline, thus making the stories comparable. Time is one of the most influential foci in that the political goals and practices of a party are particularly dependent on the time span of their ideas when climate is the issue. Because of the prominence of this aspect, and to narrow my focus due to space limitations, time will be the subject of my further analysis of climate communication in the political system.

Parties that prioritise, for example climate awareness in their energy policy work put a long time span on the process of freeing Denmark from its dependency on fossil fuel, whereas parties that prioritise awareness concerning self-sufficiency and security in their energy policy work put a shorter time span on the process and focus more on current economic needs.

In order to provide a sufficient impression of the differences and coherency of each political storyline, the following section offers a brief yet dense presentation of how each party focuses on time.

---

7 I use the terms left and right wing well knowing that they are too general and somewhat dated, but the aim of this paper is not to come up with new or better terms.

8 Established in 1989, Red-Green Alliance, the party farthest left in Parliament, grew out of Marxist/Leninist groups. The Socialist People's Party, founded in 1959, is an offshoot of the Communist Party of Denmark and calls itself a modern left-wing party. Founded in 1871, the Social Democrats, the old Labour Party, has working class roots and is close to the political centre. The Danish Social Liberal Party, established in 1905, is an offshoot of Venstre, the old Liberal Party. Currently (and during the period being studied), the Conservative People's Party, founded in 1915 as an offshoot of the old Conservative Party, Højre (Right), forms the government together with Venstre, which was established in 1870 in opposition to Højre and has roots in agriculture. Liberal Alliance, a new liberal party established in 2007, focuses on deregulation and lower income taxes. The Danish People's Party, founded in 1995, is exceedingly nationalistic and the party farthest to the right in Parliament.

Time

Red-Green Alliance

The Red-Green Alliance spokesperson states that industrialised nations have to take responsibility for the abuse by earlier generation of resources and take into account not only the opportunities open to coming generations but the climate debt owed to the Third World. He also asserts that policies that only take industrialised societies into account and that only consider societies as they are currently taking shape will give a slightly different political result.

Jointly with the remaining opposition, the Red-Green Alliance has completed and launched a motion on how Denmark can become independent of fossil fuels by 2050. The party wants Denmark to be a frontrunner in reducing CO_2 emissions regardless of the action other countries take.

Taking responsibility for earlier and coming generations is why global justice is an important part of the party's policy and why independence from fossil fuels should become a permanent part of political decision making. This is also the reason why Red-Green Alliance underlines how the superficial agreement presented in the media and often mentioned by right-wing politicians hides the lack of will to act and how weak the political agreement actually is.

Socialist People's Party

As part of the opposition the Socialist People's Party took part in the motion on freeing Denmark from dependency on fossil fuels by 2050 and on the Danish power supply becoming independent as early as 2035. The party spokesperson emphasises that calculating backwards is necessary to achieve these goals. She also states that this would call for e.g. taking traffic and agriculture policy into consideration as well as additional investment in renewable energy.

Climate and environmental protection have been an important high-priority policy area for the party for many years, not only as a separate policy, but also as an integral part of the party's other political priorities. Less than six months after COP15 and despite broad political agreement in the Danish Parliament that the climate is one of the greatest challenges of our time, the spokesperson explains that any political effort regarding the climate must incorporate financial arguments. The climate alone is no longer a sufficient argument. Taking more of a financial approach influences the rhetoric of every party, including the Socialist People's Party. The spokesperson also looks at it from another angle by mentioning how traditional ways of thinking about the climate as a long list of what to ban must be abandoned and seeing the climate as an opportunity must be embraced.

Social Democrats

As part of the opposition the Social Democrats also took part in the motion on Denmark becoming independent of fossil fuels by 2050, a stance that fits well with the political aims stated by the party's spokesperson.

Social Democrats believe that dealing with the climate means being aware of our responsibility and considering what is left behind for future generations. The spokesperson states that waiting is not enough, because the problem will not disappear by itself. This is why the party finds that going beyond its obligations, acting now and looking at mitigation and technology in the field of renewable energy are viable options. The spokesperson underlines the importance of seeing these climate strategies as political decisions and not only as new technologies and innovative companies.

Danish Social Liberal Party

The Danish Social Liberal Party was also part of the opposition in the motion for Denmark to become independent of fossil fuels by 2050. The spokesperson points out that the opposition sees being an international frontrunner in the area of climate technology and research as more important than having Denmark as an international leader in the area of climate technology prioritised by the government. Being a trendsetter in this area, according to the Danish Social Liberal Party, would create an important source of income for Denmark in the future.

The Danish Social Liberal Party maintains that climate efforts usually have high costs initially, but are often profitable in the long run. The climate spokesperson states that for the government, the political struggle does not involve prioritising the climate very highly. The party, which believes that an ounce of prevention is worth a pound of cure, considers having an international climate policy that prevents future problems instead of spending more money cleaning up catastrophes to be the most necessary and smartest solution.

Conservative People's Party

The Conservative People's Party agrees on the goal of no longer being dependent on fossil fuels because they want to be as self-sufficient as possible and reduce CO_2 emissions due to climate change. The party would like to achieve these goals at a reasonable pace, but not, according to the spokesperson, at a speed that accelerates the process to the point of becoming less competitive, adding that it will never be financially profitable to proceed too quickly.

Venstre

The spokesperson for Venstre points out how a schism exists between reducing CO_2 emissions and maintaining viable agricultural businesses. Consequently the party takes economic growth into consideration in negotiations on reducing CO_2 emissions. Venstre also considers the financial costs involved, stating that only few economies currently have the wherewithal to manage due to the financial crisis. When the party emphasises the level of political agreement on reaching the goal of preventing global temperatures from rising a maximum of two degrees Celsius above pre-industrial levels, they say that it is the timeframe that causes political disagreement.

Venstre views Denmark as a frontrunner in renewable energy technology and wants to maintain this position. The party sees renewable energy as an important, progressive sector for fulfilling their goal of creating new jobs. Nonetheless, Venstre does not advocate Denmark as a leader when it comes to CO_2 emissions, because this should be an international issue involving joint goals and agreements.

The party wants to reduce CO_2 emissions, but based on other political priorities and not with Denmark as a pioneer. How to approach this dilemma is prominent in the party's political storyline and to some extent reflects internal disagreement about climate policy within Venstre.

Liberal Alliance

The Liberal Alliance is heavily concerned about the tendency of climate policy to have far-reaching time spans. According to the spokesperson, it is impossible to say anything sensible about what is going to happen politically, technologically and climatically in 40 years' time. He explains that looking back 40 years ago allows us to understand why politicians at that time were unable to make policies suitable for use today. According to this standpoint, he argues that the present financial world should not be limited by expectations concerning climate changes in the future as this would also limit the development of new technology and the renewable energy sector.

According to Liberal Alliance the only way to make the use of energy carbon neutral is to install nuclear power stations in Denmark. The opinion is that even if there are five times as many wind turbines, twenty times as much solar power and every type of biomass available is also exploited, only 20% of the county's energy needs will be met. According to the spokesperson even these goals are highly unrealistic. The party believes that the only solution is nuclear energy, if Denmark is to become 80% carbon neutral. The goal of the Liberal Alliance is to expose the supposed charade taking place in which the government and others try to convince people that becoming independent of fossil energy is possible.

Danish People's Party

The Danish People's Party does not believe CO_2 emissions cause climate change. The party spokesperson states that only 0.032% of the atmosphere is CO_2 and that according Danish scientist Jørgen Peter Steffensen only 5% of that is human made. The spokesman claims that these results are indisputable. The party contends that such a minor amount of CO_2 cannot have a decisive effect on climate change. The Danish People's Party wants the Earth to be clean, but does not believe humans can alter the climate, the spokesperson stating that the party of course wants to minimise damage. The spokesperson also emphasises that CO^2 emissions are not bound by national borders, which is why the party finds the demand for Danish enterprises to reduce CO^2 output unreasonable if it means they move to other countries.

As a result the party currently has no intention or specific motivation to reduce CO^2 emissions based on arguments about the climate. The party promotes nuclear energy and renewable energy technology in order to become independent of the oil countries, but it does not believe this will prevent climate change, which they believe is a naturally occurring phenomenon.

The storylines of the different parties reveal the obvious differences in how they each understanding the climate. There is some agreement and some disagreement, but each party has a coherent storyline and understanding specific to its own party. In that sense, the storylines represent each party's categorisation of the climate and thus as the marked side of the distinction climate / not climate. The unmarked side is what is not climate, i.e. that which can be seen as the other parties' stories. Especially the storylines of the oppositional parties will be on the unmarked side, or not climate, as a relatively high level of internal agreement exists in both the left and right wing, as well as the greatest amount of disagreement and difference. In that sense the discussion is typical of any discussion involving the two wings and comprises environmental political priorities similar to what the parties have followed for many years. The variety of stances differ, however in nature from the quite broad agreement in the Danish Parliament that the climate challenge requires action and establishing international agreements as quickly as possible, a path clearly endorsed in 2009 with the culmination of COP15.

The semantic analysis of the documents shows that the understanding of the climate is either one of *mitigation* or *adaptation*, two aspects closely tied to the time span. Mitigation is understood as a long-term strategy, while adaptation is understood as a short-term, perhaps ad hoc strategy. Mitigation is expected to be expensive and difficult to achieve, while adaptation is expected to be cheaper and easier to achieve. Every single party communicates both strategies in their political stories, but weighs each strategy differently. This schism is the focus of the form analysis.

The Form Analysis of Climate Communication in the Danish Parliament

The form analysis is based on a distinction between unit and difference (unit / difference). As explained above, a distinction is drawn from a blind spot and one of the aims of the form is to find the distinction between unit and difference and the blind spot from which it is drawn. The blind spot of this specific distinction is what Luhmann calls the dominant form of communication (Åkerstrøm Andersen 1999). The following graph illustrates the distinction and blind spot as applied by Luhmann, who was inspired by mathematician George Spencer Brown:

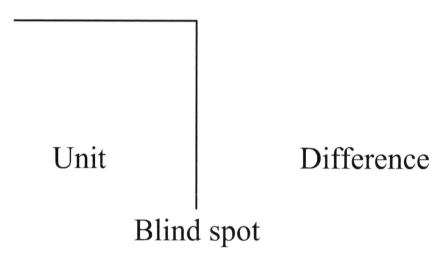

Unit   Difference

Blind spot

Figure 4.5.1 Dominant form of communication

The distinction of unit and difference are paradoxical in nature, which means there is a distinction between the two sides, which are, at one and the same time, impossible to connect, but in connecting, they motivate further action and communication. By observing and analysing climate communication in the Danish Parliament and by constructing the political storylines of the different parties, it becomes obvious that the distinction between unit and difference in the political system is the distinction between *mitigation* and *adaptation* – and thus the distinction between a long time span and a short one. This distinction is the one that unites the communication of all of the different politicians and political parties on climate change.

The relationship between *mitigation* and *adaptation* is of paradoxical nature because it is difficult to connect the two in political decisions as every single decision on climate action will be a decision involving either mitigation or adaptation, which means that the decision will always remain solely on one side of the distinction. At the same time, connecting these two types of decisions on climate action sets the agenda of the communication and subsequent decisions made in the political system. This means that mitigation and adaptation are impossible to combine in the same political decision but are at the same time related in that they force the agenda for further decisions made on the climate in the political system.

Mitigation is articulated as a long-term strategy that is expensive and difficult to achieve, while adaptation is articulated as a short-term strategy that is cheaper and easier to achieve. From a political perspective, the tendency is for right-wing parties to talk about climate and mitigation as currently being an unaffordable financial cost and, in light of the financial crisis, to claim that the climate should not be prioritised as one of the most important political issues. Prioritising the climate right now will further harm Danish business conditions. Left-wing parties, on the other hand talk about climate as a social cost because not prioritising it high enough in the beginning will mean additional costs for industry and trade, especially regarding renewable energy technology.[9] Thus left-wing parties see a strategy based on adaptation as a social cost because adaptation means not prioritising the climate. Cost understood as both financial costs and social costs seems to be of importance in the political communication on the climate. Cost is what steers how adaptation and mitigation are prioritised and which actions are easily achievable or not.

By observing and analysing climate communication in the Danish Parliament and constructing political storylines, the blind spot then becomes obvious and consequently the dominant form of communication in the communicative practice regarding the climate. This dominant form of communication is the *cost for society*, which is in turn the blind spot from where the distinction between adaptation and mitigation is drawn. Cost for society is dominant in that it is the form that all of the communication about the climate is referred to. It is the unit of all of the distinctions.

Thus the dominant form of communication on how the climate is understood in the Danish Parliament is apparently the cost for society. The dominant form of communication can be illustrated as Figure 4.5.2.

Overall political climate action shows that it is of course possible to see decisions about both mitigation and adaptation, but every decision is still either or. The political storylines of the parties comprise a paradoxical relationship between mitigation and adaptation as every party, over time, decides or believes in both mitigation and adaptation. As illustrated by the political storylines presented, big differences exist between how the parties weight mitigation and adaptation. Left-wing parties clearly weight mitigation with a long time span higher than adaptation with a short time span, whereas right-wing parties weight adaptation higher than mitigation. How the parties prioritise is of course interconnected with how they perceive the costs for society.

---

9   The Socialist People's Party spokesperson also points out that 2009 was a year in which there was a political opportunity to talk about the climate and focus on climate policy. As of May 2010, in contrast political climate discussions and proposals now have to be dressed up in economic arguments that must be financially beneficial in either the short or long run.

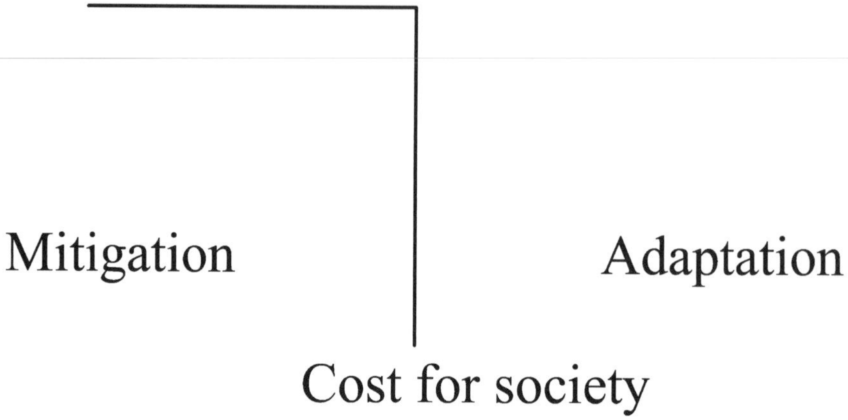

Mitigation | Adaptation

Cost for society

Figure 4.5.2 Dominant form of communication in the Danish Parliament

Communication Praxis About Climate in the Political System

Cost for society appears to be the dominant form of communication when politicians in the Danish Parliament communicate about the climate as well as the form of communication in which communication in general is connected. Furthermore the distinction between adaptation and mitigation is the distinctive communication that forces communication about the climate in the political system to continue.

With regard to understanding society as differentiated, this specific communication praxis about climate should be seen as system specific and thus as operating within the binary code and logic of the political system. Therefore both communication about cost for society and the distinction between mitigation and adaptation can primarily be understood as an internal political affair whose efforts are to gain voters and political power. In that sense the distinction and discussion on mitigation and adaptation and the discussion about cost for society in the political system can be better grasped looked at as a discussion and distinction stemming from the different political wings, which obviously show voters the difference between the political parties, than it can be grasped as stemming from a political will to solve climate change issues. This is what Luhmann describes as the political system's limited resonance of ecological problems (1989).

The system-specific understanding and communicative praxis about the climate and the limited resonance of climate changes based on the premises of the environment or climate threats further underline how every system's communication and concerns about climate change determine how climate issues are constructed and thus adapted into irritations about the climate. In that sense every acknowledgement of the climate is adaptation and the distinction between adaptation and mitigation appears to be an artificial one, but apparently a politically functional distinction, because mitigation cannot exclude adaptation but is itself adaptation.

When the cost for society is the dominant form of communication about climate in the Danish Parliament, it is both an opportunity and a barrier. It is an opportunity, because it eases communication about the climate as a cost for society then becomes the framework of climate communication in general in the Danish Parliament. It is also a barrier, because this specific form and framework limits communication about climate. For example climate communication as a cost for society excludes communication about the *climate as an opportunity*. Only a few politicians try to communicate climate as an opportunity, but if this approach was the dominant form of communication in the Danish Parliament, the force of the distinction of the communication would probably be different than the distinction between adaptation and mitigation. Of course it would still be based on the premises of the political system, but somehow make a different agenda for climate communication in the political system.

Final Remarks

The importance of this sort of close-up analysis is to show how climate is communicated, negotiated and understood in the political system in Denmark surrounding COP15 in December 2009. The goal was to see if the apparently broad agreement among Danish politicians also proved to be the case if the communications and the negotiations that took place were investigated more closely.

The proposals and negotiations transcribed in *Folketingets Tidende* reveal a high level of agreement in the topics surrounding and definitions of climate among the different parties. The disagreements visible in the negotiations are usually about the means of achieving the common goal of a Denmark independent of fossil fuels and the discussion usually digresses from the climate and takes shape as a discussion of well-known specific disagreements between the parties.

Interviews with the climate spokespeople show that the political discussion follows a traditional pattern of a high level of disagreement between the right and left wing. The right-wing politicians understood climate as financial cost and as such a problem for the Danish business environment. Interconnected with this is how they weight adaptation higher than mitigation because the initiation of a long-running climate policy would be a cost upon its initiation. Left-wing politicians, on the other hand understood climate as a social cost, and if not prioritised highly, would represent a cost to industry and trade, especially in the field of renewable energy technology. Interconnected with this is how they weight mitigation higher than adaptation because initiating a long-running climate policy would save additional social costs arising from climate change and give the climate business far better conditions in the future than any short-sighted economic benefits gained from not initiating a long-term climate policy freeing Denmark from its dependency on fossil fuels.

Consequently, disagreement on the issue is vast among the parties, even though a level of agreement is apparent when they talk about the climate as a cost for society as the dominant form of communication in the communication about the climate in the Danish Parliament. Even when some of the parties mention climate as a political opportunity the dominant form of communication as a cost for society is a strong barrier against seeing the climate as an opportunity for economic or societal change as proposed, for example by British economist Nicholas Stern (2009).

Awareness and acceptance of society as differentiated is an important step towards understanding the difficulty of handling climate change, because it creates conscious discernment concerning society's differentiated and seemingly limited resonance of climate change. When the distinction between adaptation and mitigation and the dominant form of communication is cost for society are the resonance of climate change in the political system, the political system could, with the acceptance of a differentiated society, become aware of this limitation and perhaps work to change this dominant form of communication. This change would still be based on political premises, but possibly lead to a more successful handling of climate change than currently observable in the political system. This means that it is important to investigate and understand more than just the political system in order to handle climate change, but political systems are and will continue to be, as a home base for communication on ecological problems, of particular importance to understanding and creating irritation. Producing irritation of this kind is one of the goals of presenting the analysis undertaken in this chapter.

References

Åkerstrøm Andersen, N. 1999. Diskursive analysestrategier. Foucault, Koselleck, Laclau, Luhmann. Copenhagen. Nyt fra Samfundsvidenskaberne.

Almlund, P. 2007. Miljøkommunikation i virksomheder – praksis i kontekstens blinde plet. Ph.D. Thesis. Department of Communication, Business and Information Technologies. Roskilde University, Denmark.

Bourdieu, P. 2005. Viden om viden og refleksivitet. Forelæsninger på Collège de France 2000–2001. Copenhagen. Hans Reitzels Publishers.

Bourdieu, P. 1997. Af praktiske grunde. Omkring teorien om menneskelig handlen. Copenhagen. Hans Reitzels Publishers.

Bourdieu, P. and L. J. D. Wacquant 1996. Refleksiv sociologi. Hans Reitzels Publishers.

Bourdieu, P. 1977. Outline of a theory of practice. United Kingdom. Cambridge University Press.

Callewaert, S. 1997–1998: Bourdieu-studier. Institut for Filosofi, Pædagogik og Retorik. Københavns Universitet Amager, Vol. 1–3. Copenhagen.

Kneer, G. and A. Nassehi 1997. Niklas Luhmann – introduction til teorien om sociale systemer. Copenhagen. Hans Reitzels Publishers.

Luhmann, N. 2002: Massemediernes realitet. Copenhagen. Hans Reitzels Publishers.

Luhmann, N. 2000. Sociale systemer. Grundrids til en almen teori. Copenhagen. Hans Reitzels Publishers.

Luhmann, N. 1998. "Erkendelse som konstruktion" in M. Hermansen (ed.) Fra læringens horisont – en antologi. Århus. Klim.

Luhmann, N. 1997. Iagttagelse og Paradoks. Copenhagen. Gyldendal.

Luhmann, N. 1995. "The paradoxy of observing System", Cultural Critique, no. 31, p. 37–55.

Luhmann, N. 1992. "Operational closure and structural coupling: The differentiation of the legal system", Cardozo law review Vol. 13 s. 1419–1441.

Luhmann, N. 1989. Ecological Communication. Great Britain. Polity Press.

Stern, N. 2009. His speech at the Conference "Climate Change. Global Risks, Challenges & Decisions", Copenhagen 2009, 10–12 March.

Thyssen, O. 1997. "Hjørnesten i Niklas Luhmanns systemteori" in Niklas Luhmann 1997: Iagttagelse og Paradoks. Copenhagen. Gyldendal.

Index